How to Solve
Word Problems in Mathematics

David S. Wayne, Ph.D.

McGraw-Hill

New York   San Francisco   Washington, D.C.   Auckland   Bogotá
Caracas   Lisbon   London   Madrid   Mexico City   Milan
Montreal   New Delhi   San Juan   Singapore
Sydney   Tokyo   Toronto

Library of Congress Cataloging-in-Publication Data applied for.

McGraw-Hill

A Division of The McGraw·Hill Companies

1 2 3 4 5 6 7 8 9 0   DOC/DOC   0 9 8 7 6 5 4 3 2 1 0

ISBN 0-07-136272-X

The editing supervisor was Maureen B. Walker and the production supervisor was Tina Cameron. It was set in Stone Serif by PRD Group.

Printed and bound by R. R. Donnelley & Sons Company.

McGraw-Hill books are available at special quantity discounts to use as premiums and sales promotions, or for use in corporate training programs. For more information, please write to the Director of Special Sales, Professional Publishing, McGraw-Hill, Two Penn Plaza, New York, NY 10121-2298. Or contact your local bookstore.

This book is printed on recycled, acid-free paper containing a minimum of 50% recycled, de-inked fiber.

Contents

Preface

In nearly every part of the United States, students in grades 6 to 12 are undergoing statewide assessments in mathematics that differ significantly from tests that were taken prior to the late 1990s. These current tests are based on revised standards that were created to address a noticeable nationwide weakness in applying the principles and computational skills in mathematics to realistic, relevant, and practical situations. Most state curriculum documents list as a key component mathematical reasoning and modeling. These are reflected in the assessments in the form of word problems, and, in most cases, students are required to not only produce a correct answer but also to carefully *explain* their work. Additionally, many of these assessment tests cover multiple grade levels and require students to recall facts and procedures that may not be part of their current year's course.

This book is designed to enhance the student's ability to recall those facts and procedures that are necessary for success in solving word problems. The problems offer opportunities to review the important areas of measurement, estimation, formula application, algebraic representation, interpreting geometric situations, data analysis, and probability. The solutions include strategies for decoding problems, determining what skills and concepts need to be applied, and how to present a solution that shows complete understanding. This book is not designed as a textbook that presents information to be learned for the first time; however, there is sufficient explanation of key concepts to enhance the student's understanding. There is also an appendix which contains a review of solving linear and quadratic equations. Each chapter in the book focuses on word problems that have a common element.

Chapter 1 presents problems involving measurement and the application of simple formulas. The information and concepts used in these word problems form the core of most problems that students face, and these problems are solved by simple arithmetic and reasoning.

Chapter 2 helps the students solve more complex problems using algebraic skills and modeling. This chapter provides review and practice in solving linear and quadratic equations.

Chapter 3 focuses on problems for which students need to recognize the use of ratios and proportions and apply them correctly.

Chapter 4 presents the student with word problems that involve geometric relationships and ways to model these situations.

Chapter 5 presents examples of interpreting data and using the basic principles of probability and statistics to respond to questions about the data.

Chapter 6 is a collection of problems that are not categorized and provides additional practice for students in determining how these problems can best be solved.

Whether used in a classroom setting with guidance by a teacher or at home as a supplement to class, this book will provide students with opportunities to better prepare them for being successful in applying mathematics.

I'd like to take this opportunity to thank my wife, Phyllis; my daughter, Gayle; my friends and mentors, Carl Goodman and Bert Linder; and many other family members, colleagues, and friends. Without their support and encouragement, this endeavor would not have been possible.

Chapter I

Measurement, Estimation, and Using Formulas

You have probably found solving word problems to be the most difficult part of any math course. In most cases, these involve applications of well-known formulas and an understanding of the relationships between quantities. The key idea in solving these problems is to identify the way in which the quantities are being measured and to recognize which formula to apply. It is also helpful to be able to estimate an answer so that you can determine whether your worked-out solution is reasonable.

The examples that follow will show you how to use some of the most common formulas and measurements that commonly appear in word problems. All the formulas and measurements are listed at the end of this chapter for reference.

When measurements and calculations result in numbers that are not whole and have long decimal parts, it makes sense to round answers to one, two, or three decimal places depending on the accuracy you need in order to solve the problem. Sometimes a problem will clearly tell you how many decimal places to use. For example, "Find the length to the *nearest foot*" requires you to give your answer rounded to the nearest whole number. Your calculations should be performed using at least one decimal place to ensure that your answer is rounded correctly. In general, you should use at least *one more decimal place* in your calculations than the answer requires. If the problem does not specify how you should state your answer, you should use a reasonable amount of decimal places so that your final answer will be meaningful and practical. For example, when a problem involves money, you could use two decimal places since we rarely calculate with fractions of a penny. Therefore, it is important to look for how an answer is to be given and to think about what would be a reasonable answer.

Distance

Measuring distance is the same as measuring length. In the English (i.e., British and U.S. Customary) system, the units of measuring distance include inches, feet, yards, and miles. In the metric system [i.e., International System of Units (SI)], the units include millimeters, centimeters, meters, and kilometers. It is important to know how one unit of measurement can be converted to another. An easy way to convert measurements is to divide the number of smaller units by the amount of those units that are in the larger unit. For example, the number of feet (ft) is equal to the number of inches (in) divided by 12, or the number of meters (m) is equal to the number of millimeters (mm) divided by 1000. For conversions that aren't as simple, another method will be shown.

An important measurement of distance is *perimeter*, which is the distance around a region. This is used in the example below.

Example I

A rectangular room measures 216 in by 150 in. The interior designer wants to place a strip of molding around the room near the ceiling that costs $2.25 per foot. How much will it cost to buy the molding?

Solution I

The answer lies in finding the perimeter of the room first in inches and then in feet. The perimeter is the sum of all the edges of the shape. In the case of a rectangle

$$\text{Perimeter} = 2 \times \text{length} + 2 \times \text{width}$$

2

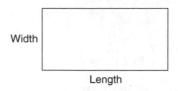

Width

Length

Therefore, the perimeter of our room is $2 \times 216 + 2 \times 150 = 732$ in. We convert this to feet by dividing by 12, and we have 61 ft. The cost of the molding would be

$$61 \text{ ft} \times 2.25 \frac{\text{dollars}}{\text{foot}} = \$137.25$$

Note that the units in this product are written as fractions. This is a good device to help determine whether the items being multiplied are the correct ones and are in the correct order. The units can be multiplied as if they were fractions and, after "canceling" similar units that appear in the numerators and denominators, the resulting fraction should be the units you wish your answer to have. Thus, for this problem, (feet/1) × (dollars/foot) = (dollars/1) = dollars.

Example 2

While planning a trip around Italy, Jim wanted to travel from Rome to Milan to Venice and back to Rome. On a map he found that the distances were approximately 300 kilometers (km) from Rome to Milan, 150 km from Milan to Venice, and 225 km from Venice to Rome. Jim was not comfortable with these measurements and wanted to know the distance in miles. To the nearest mile, what is the total distance of the trip?

Solution 2

The conversion for miles to kilometers in the reference table states that 1 mile (mi) is 1.61 km. If we think of this as a fraction, we have

$$\frac{1 \text{ mi}}{1.61 \text{ km}} = 0.62 \text{ mi/km}$$

For our problem, we need to add the distances in kilometers, $300 + 150 + 225 = 675$, and then multiply it by 0.62. The

3

answer is

$$675 \, km \times 0.62 \, \frac{miles}{kilometer} = 418.5 \, mi = 419 \, mi$$

Remember that an estimate is always useful to have first to be sure that your answer makes sense. Since the conversion factor is greater than 0.5, you would know that your answer would have to be more than ½ of $300 + 150 + 225$ or ½ of $675 = 337.5$ mi and you would feel comfortable with the answer of 419.

Distance traveled can also be measured indirectly by knowing the speed at which the travel was occurring and the time taken for the journey. The distance is computed by the simple formula *distance = rate × time*.

Example 3

A car has been traveling at an average speed of 55 mph (miles per hour) for 7 hours. How many miles has the car traveled?

Solution 3

Note the phrase "average speed." The car may have traveled at somewhat different speeds throughout the 7 hours. However, for the purposes of this problem, you can assume that the car has been traveling at that average speed throughout the journey. (When making this assumption, we usually say that the car has traveled at a *constant speed*.)

The question "How many miles...traveled?" is really asking "What is the total distance traveled?" Since you know the rate and the time, you can use the formula directly:

$$Distance = 55 \, mph \times 7 \, hours = 385 \, mi$$

Notice that mph is really the ratio miles/hour, and we are once again multiplying units to arrive at our answer in miles.

Area

Area is the measurement of the two-dimensional space within a region such as the space on the floor of a room, the space on a canvas of a picture, or the space on which a house is built.

4

Note that the answer to an area problem uses square units. This leads to conversions different from those in measuring one-dimensional length.

$$1 \text{ square foot (ft}^2) = 12 \text{ in} \times 12 \text{ in} = 144 \text{ square inches (in}^2)$$

This means that within a square that measures 1 ft × 1 ft, you can fit 144 squares that measure 1 in by 1 in.

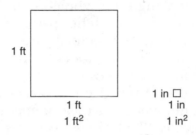

$$1 \text{ square meter (m}^2) = 100 \text{ centimeters (cm)} \times 100 \text{ cm}$$
$$= 10,000 \text{ square centimeters (cm}^2)$$

This means that within a square that measures 1 m by 1 m, you can fit 10,000 squares that measure 1 cm by 1 cm.

When measuring area, it is important to make sure that the two measurements that are being multiplied are in the same units. This would require making conversions as we did when measuring length.

Example 4

A rectangular plot of land is roped off for planting. The measurements are 8 ft 4 in by 5 ft 8 in. Each plant requires 1 square yard (yd^2) of space for planting. How many square yards have been roped off? How many plants can be planted in this plot of land?

Solution 4

Be sure to draw a rectangle and label the sides with the given measurements! Drawing diagrams is an important good habit.

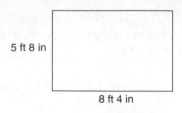

5 ft 8 in

8 ft 4 in

In this problem it is important to convert all measurements to the same units. Since we don't plant fractions of a plant, we will be looking for a whole-number answer. Therefore, we should convert the inches parts of our numbers to their equivalent fractional or decimal parts of a foot. Thus, for example, 4 in is one-third of a foot (1/3 ft) or 0.33, and 8 in is two-thirds of a foot (2/3 ft) or 0.66. We can also round our measurements to one or two decimal places since our answer must be in whole numbers. The measurements become 8.33 ft by 5.67 ft. The area is 47.23 square feet or ft^2. Since 1 square yard (yd^2) is 3 ft by 3 ft, there are 9 ft^2 in 1 yd^2. There are approximately

$$8.33\,\text{ft} \times 5.67\,\text{ft} \times \frac{1\,\text{yd}^2}{9\,\text{ft}^2} = 5.25\,\text{yd}^2$$

in this plot of land.

The answer to the second question is a little trickier. If each plant truly requires a square shape in which to be planted, then we have to look at a diagram to see that there is enough room for only two plants!

5.67 ft

Plant

8.33 ft

Plant

This points out that when we measure area, we are not actually counting squares. The quantity that we call the *area* suggests that if we could rearrange the space, we could construct that number of squares. The preceding diagram shows that there are only two actual squares in this plot of land that are 3 ft by 3 ft. The rest of the space could create an additional 3.25 squares *if it were possible*.

Circles are popular in word problems of this type. The distance around the circle is referred to as its *circumference* (rather than perimeter), and the space within the circle is its area. The values of these measurements involve the number π (Greek lowercase pi), which doesn't have an exact value, but we approximate it with either 3.14 or with the fraction 22/7. (Technical work, such as engineering, often requires better approximations. Supercomputers have calculated π to billions of decimal places beyond 3.14.) The diameter and radius of the circle are also involved in the calculation.

The formulas are circumference $= \pi \times$ diameter or $2 \times \pi \times$ radius, because the diameter is equal to twice the radius, and area $= \pi \times$ radius^2.

Example 5

An interior designer wants to cover the floor of a room with a carpet that has a two-color design. The room is square, measuring 12 ft on each side. The designer wants a red circle, whose center is the center of the room, that leaves 2 ft to each of the four walls; the remainder of the carpet should be white (i.e., there should be a 2-ft white-carpet border around the red-carpet circle in the center of the room). The red carpet costs $12 per square foot, and the white carpet costs $10 per square foot. How much will the carpeting cost?

Solution 5

Once again, it is useful to estimate. The entire room is 12 ft by 12 ft, which creates an area of 144 ft^2. Using the two prices, the cost should be between 144 ft$^2 \times \$10/\text{ft}^2 = \$1440$ and 144 ft$^2 \times \$12/\text{ft}^2 = \$1728$. (If this were a multiple-choice question, you might find that only one of the choices falls

7

into this range. That would have to be the correct choice, and your work would be done!)

A diagram is required to find the actual answer.

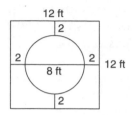

Since we want to leave 2 ft to the wall, the diameter of the circle must be 8 ft and, therefore, its radius must be 4 ft. We use the formula for the area of a circle, area $= \pi \times \text{radius}^2$, and find that the red circular area would be approximately $3.14 \times 4^2 = 50.24\,\text{ft}^2$. The cost of the red carpet would be $50.24\,\text{ft}^2 \times \$12/\text{ft}^2 = \$602.88$.

The remaining area in the room would be $144\,\text{ft}^2 - 50.24\,\text{ft}^2 = 93.76\,\text{ft}^2$, and the white carpet would cost $93.76\,\text{ft}^2 \times \$10/\text{ft}^2 = \$937.60$.

The total cost would be $1540.48.

There is a measurement of land area that doesn't directly state that there are square units, but it is implied. This is referred to as an *acre* ($1\ \text{acre} = 43{,}560\,\text{ft}^2$).

Example 6

A farmer has a farm $\frac{1}{2}$ mi $\times 1\frac{1}{4}$ mi. How many acres is this farm? How many different crops can the farmer plant, if each crop requires $\frac{1}{2}$ acre?

Solution 6

The farm has $0.5 \times 1.25 = 0.625\,\text{mi}^2$. We need to convert this to acres. The conversion given above relates acres to square feet. We can set up a multiplication of fractions using the correct units in the numerators and denominators to arrive at the number of acres on the farm.

$$\frac{0.625\,\text{mi}^2}{1} \times \frac{5280^2\,\text{ft}^2}{\text{mi}^2} \times \frac{1\,\text{acre}}{43,560\,\text{ft}^2} = 400\,\text{acres}$$

The farmer could plant $400 \div \frac{1}{2} = 800$ crops.

Volume

Volume is a measure of the amount of material a three-dimensional object can hold. The most common objects for which we use formulas are rectangular boxes, circular cones, and spheres. The formulas are given in the reference table at the end of the chapter under the subheading "Measuring Volume."

Note that the units of volume are cubic units since we are multiplying three measurements. A *cube* is a rectangular box whose length, width, and height are all equal.

$$1\,\text{ft}^3 = 12\,\text{in} \times 12\,\text{in} \times 12\,\text{in} = 1728\,\text{in}^3$$

This means that into a box that measures 1 ft by 1 ft by 1 ft, we can fit 1728 boxes that measure 1 in by 1 in by 1 in.

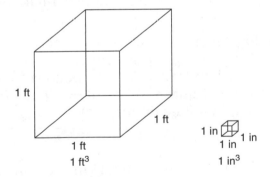

Similarly, 1 cubic meter (m^3) = 100 cm × 100 cm × 100 cm = 1,000,000 cubic centimeters (cm^3). This means that into a box that measures 1 m by 1 m by 1 m, we can fit 1 million boxes that measure 1 cm by 1 cm by 1 cm.

Example 7

Paper is shredded into rectangular bins that measure 24 in by 12 in by 12 in. These bins are emptied into larger cubical bins that are 4 ft on each side. How many smaller bins can be emptied into each cubical bin?

Solution 7

Since we are interested in knowing what it takes to fill something, we need to measure volume. If we can determine the volume of each smaller bin and the volume of the cubical bin, we can divide the volume of the cubical bin by the volume of each smaller bin to arrive at our answer.

The volume of each smaller bin is $24 \text{ in} \times 12 \text{ in} \times 12 \text{ in}$ or $2 \text{ ft} \times 1 \text{ ft} \times 1 \text{ ft} = 2 \text{ ft}^3$. The volume of the cubical bin is $4 \text{ ft} \times 4 \text{ ft} \times 4 \text{ ft} = 64 \text{ ft}^3$. Therefore, we can empty $64 \div 2 = 32$ smaller bins into the larger.

We often use different units to measure the volume of a liquid. These don't seem to be cubic units, but they really are. For example, in the metric system we use *liters* to measure volume. We use the following conversion formula to understand how liters are related to cubic units.

$$1 \text{ milliliter(mL)} = 1 \text{ cubic centimeter (cm}^3 \text{ or cc)}$$

This means that $1 \text{ liter (L)} = 1000 \text{ cm}^3$.

The English system of measurement uses the gallon as its standard, and $1 \text{ gallon (gal)} = 231 \text{ in}^3$. Gallons are subdivided as follows: 1 gal = 4 quarts (qt), 1 qt = 2 pints (pt), 1 pt = 2 cups, and 1 cup = 4 fluid ounces (oz).

Example 8

A rectangular fishtank is 30 in long, 15 in wide, and 20 in high. Gravel is laid evenly at the bottom of the tank to a height of 2 in, and the tank should be filled to a point that is 3 in from the top of the tank. How much water can the tank hold to the nearest tenth of a gallon?

10

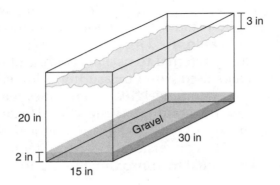

After the gravel is laid, the height available for water is $20 - (2 + 3) = 15$ in. Therefore, the tank can hold $30 \, \text{in} \times 15 \, \text{in} \times 15 \, \text{in} = 6750 \, \text{in}^3$ of water. Our answer, however, should be in liquid measurements.

$$\frac{6750 \, \text{in}^3}{1} \times \frac{1 \, \text{gal}}{231 \, \text{in}^3} = 29.22 = 29.2 \, \text{gal}$$

Example 9

A balloon when expanded is in the shape of a sphere with a diameter of 24 cm. What is the maximum capacity of the balloon to the nearest liter?

Solution 9

Capacity refers to the amount an object can hold; therefore, we need to measure its volume. The reference table at the end of this chapter shows that the formula for the volume of a sphere is given in terms of its radius, which is one-half of the diameter, which in this case would be 12 cm. Using the formula for the volume of a sphere and 3.14 for π, we calculate the volume to be $4/3 \times 3.14 \times 12^3 = 7234.56 \, \text{cm}^3$. This is equivalent to 7234.56 milliliters (mL) and, dividing by 1000, to 7.23456 L. Our answer would be 7 L.

Weight

Weight is also measured in both the English and metric systems in different ways. In the English system, we use the ounce as the smallest standard unit and the pound (lb; 1 lb = 16 oz) and ton (2000 lb) for larger weights. (The ounce used for weight is usually referred to as a *dry ounce* and is different from the *fluid ounce* used when measuring volume.)

In the metric system, weight is sometimes referred to as *mass*. Here, we use the gram (g) as the smallest standard unit and the kilogram (kg; 1 kg = 1000 g) and metric ton (1000 kg) for larger weights. The conversion between systems most often used is that 1 kg is equal to approximately 2.2 lb. Also, 1 kg is the exact weight of 1 L of water.

Example 10

In order to prepare for an international wrestling event, Jim had to bring his weight to under 80 kg. If he currently weighs 200 lb and begins a program in which he loses 2.5 lb per week, in how many weeks will he reach the required weight?

Solution 10

In order to work through this problem, we need to convert the information to either pounds or kilograms. Since the rate of losing weight is given in pounds per week, we ought to convert the 80 kg to pounds. Using the "fraction" scheme of conversion, we have

$$\frac{80\,\text{kg}}{1} \times \frac{2.2\,\text{lb}}{1\,\text{kg}} = 176\,\text{lb}$$

Jim would have to lose $200 - 176 = 24$ lb. This would take $24\,\text{lb} \div 2.5\,\text{lb/week} = 9.6$ or 10 weeks to reach the eligible weight.

Example 11

The ingredients label on a box of cereal indicates that one serving of cereal contains 24 g of carbohydrates. In order to understand this better, Rachel wants to convert this to ounces. To the nearest tenth, how many ounces of carbohydrates will she be eating, if she has two servings of cereal?

We can use the "fraction" scheme to answer this question with one calculation:

$$\frac{24\,g}{1\,\text{serving}} \times \frac{1\,kg}{1000\,g} \times \frac{2.2\,lb}{1\,kg} \times \frac{16\,oz}{1\,lb} \times \frac{2\,\text{servings}}{1}$$

$$= 1.6896\ oz = 1.7\ oz$$

Temperature

Other situations that arise in word problems involve temperature. In the world today we measure temperature in two scales, Fahrenheit and Celsius (or Centigrade). The Celsius scale uses the freezing and boiling points of water to create a range from $0°$ to $100°C$. The conversion to Fahrenheit is given by the formula $F = 1.8\,C + 32$, where F represents degrees Fahrenheit and C represents degrees Celsius.

Example 12

Find the freezing and boiling points of water in degrees Fahrenheit.

Solution 12

This is a simple application of the formula.

$$\text{Freezing} - 0°C \rightarrow F = 1.8 \times 0 + 32 = 32°F$$

$$\text{Boiling} - 100°C \rightarrow F = 1.8 \times 100 + 32 = 212°F$$

Example 13

The human body maintains a normal temperature of $98.6°F$. If a person developed a fever with a $3°$ temperature increase above normal, what would be the Celsius reading on a thermometer, to the nearest tenth?

Solution 13

The new Fahrenheit temperature would be $101.6°F$. The formula can still be used, except that we would have to solve an algebraic equation to determine C.

$$101.6 = 1.8C + 32 \qquad \text{(add } -32 \text{ to both sides of the equation)}$$

$$\rightarrow 1.8C = 69.6 \qquad \text{(divide both sides of the equation by 1.8)}$$

$$\rightarrow \quad C = \frac{69.6}{1.8} = 38.66\ldots{}^{\circ}C \qquad \text{(which is } 38.7{}^{\circ}C \text{ to the nearest tenth)}$$

Shopping

Many word problems involve finding the amount one has to spend on a purchase when sales tax and/or discounts are applied. Usually we are given an initial purchase price of an item and a percentage that indicates the discount and/or tax. The solution to most of these problems is found by finding the dollar amount of those percentages and adding or subtracting. A formula one can use for finding a discount price is

Discounted price = original price − discount × original price

The discount needs to be in decimal form in order to perform the multiplication correctly. A formula for finding the cost with sales tax included is

Total cost = price + tax × price

Again, the tax should be in decimal form.

Example 14

Jim wanted to buy several compact disks (CDs). The music store was having a sale that reduced the prices by 15 percent. He selected three CDs that cost $12.99 and one that cost $9.99. How much would this purchase cost if sales tax is $8\frac{1}{2}$ percent?

Solution 14

Before figuring in the percentages, the total cost of the four CDs needs to be determined. This is $(3 \times \$12.99) + \$9.99 = \$48.96$. Using the first formula to find the new price

14

with the discount, we have

$$\text{New price} = \$48.96 - (0.15 \times \$48.96) = \$41.62$$

Note that this price could also have been calculated by finding 85 percent of the price or multiplying $48.96 by 0.85.

Using the second formula, we can determine the total cost,

$$\text{Total cost} = \$41.62 + (\$41.62 \times 0.085) = \$45.16$$

Note that this could have been calculated by finding 108.5 percent of the price or multiplying $41.62 by 1.085.

Interest

Another common application of mathematics involves finances. People are interested in determining how much money will result when interest is applied. *Interest* is simply an amount of money that is added to the original amount and calculated by a percentage of that amount. There are many complicated ways to calculate interest, but the simplest is found using what is called the *simple interest formula*:

$$\text{Interest earned} = p \times r \times t$$

where p is the original amount called the principal; r is the interest rate as a fraction, decimal, or percent; and t is the number of years that the investment is allowed to grow. The total current amount is found by adding the principal investment and the interest earned.

$$\text{Total amount} = \text{principal} + \text{interest earned}$$

Example 15

Samantha has been working hard as a baby-sitter and has $540 collected. She decides to invest this money in a fund that predicts an interest rate of 8 to 11 percent. What could she expect to happen to her original investment after one year?

15

Solution 15

You could first estimate an answer by supposing that the rate was 10 percent. This would result in Samantha receiving $54 in interest for an estimate of having a total of $594 at the end of the year. An actual solution would be as follows. At the least, Samantha would receive 8 percent interest. This would result in I = $540 × 0.08 × 1 = $43.20.

At the most, Samantha would receive 11 percent interest. This would result in I = $540 × 0.11 × 1 = $59.40. At the end of one year, using total amount = $p +$ I, she would have saved between $583.20 and $599.40.

Note that the percentage needed to be changed to a decimal in order to make the calculation. If you didn't do this, the answer would be unreasonable, as the result would be 100 times larger!

The problems that follow will give you plenty of opportunity to use the formulas and ideas that you just studied. Be sure to draw diagrams to help you see the situation clearly, and take a look at the reference table at the end of this chapter to remind yourself of the formulas.

Additional Problems

1. Gayle would like to fence her garden, which is in the shape of a triangle with two sides measuring 10½ ft and the other side measuring 8 ft. The material she chose to use costs $4.25 per meter. How much will she spend to fence her garden?

2. Gayle's triangular garden from Prob. 1 needs a plastic ground cover in order to keep weeds from growing. She measures the distance from the 8-ft side to the opposite vertex of the triangle to be 6 ft. How much plastic should she use?

3. The length around a wooden rectangular picture frame is 50 in. There is a 2-in border on all sides of the picture within the frame. If the longer side of the picture is 11 in, what is the length of the shorter side of the frame?

4. The owner of a corner store wants to paint the upper portion of a side exterior wall that is 16 ft tall to make the store more noticeable. The length of the wall is 25 ft. He buys a gallon of yellow

paint to do the job and is told that each gallon can cover 400 ft². What portion of the wall will be yellow?

5. A circular swimming pool is being installed in a neighbor's yard. The county requires that the pool be 6 ft from the end of the property. If the yard is 30 ft by 26 ft and the pool is to be centered in the center of the yard, what is the largest area that can be used for the pool?

6. Melissa is planning to have the front of her property landscaped. Her front yard is rectangular and measures 12 m by 14 m. She plans to put down sod over the entire yard, except for a brick circular design in the middle that has a radius of 10 ft. To the nearest dollar, how much will Melissa need to spend to cover the entire yard if sod sells for $6 per square foot?

7. How high would the sides of a rectangular water tank have to be if the tank base has an area of 2 m² and the tank must hold 5520 L of water?

8. An ice-cream parlor owner charges $1.95 for an ice-cream cone. The cone is first packed in, and then a scoop of ice-cream sits like a ball on top. The top of the cone has a diameter of 2 in, and the cone is 6 in tall. The scoop also has a diameter of 2 in. How much profit is the owner making, if the ice-cream costs her $30 per gallon?

9. An airplane traveling from San Fransisco to Dallas makes a stopover in Phoenix. The average speed of the plane was 320 mph. If it took 2 hours to reach Phoenix and another 2 hours and 45 minutes to reach Dallas, how many miles has the plane flown?

10. On the first day of a cross-country automobile trip with her family, Jennifer covered a distance of 680 mi in 9½ hours. The speed limit on the highway she drove is 65 mph. How much longer or shorter time would it have taken Jennifer if she drove at the speed limit?

11. The current school record for the 100-m dash is 9 seconds. At what speed, in miles per hour calculated to the nearest tenth, would a runner have to run in order to beat the speed record for the 100-m dash?

12. In a laboratory, it was found that a compound boils at a temperature of 42°C. The heating apparatus can raise the temperature by 2°F per minute. In how long a period can the compound be brought to boil if the heating process begins at 68°F?

13. While planning a winter vacation for his family, Frank investigates temperatures in different parts of the world. The information he finds on the Internet is given in degrees Celsius. He wants to travel to a place where the temperature would be between 75 and 85°F.

What range of temperatures should he enter into the search engine?

14. A store is having a "half-price sale"; that is, if you buy one item at full price, you can buy a similar item for half-price. If the sales tax is 6 percent, how much would it cost to buy two pairs of jeans, each costing $24.99?

15. An owner of a clothing store attempting to clear out her old inventory marks down the outfits she wishes to sell by 10 percent each week. Lindsey has $90 to spend on an outfit. She sees that it is currently marked at $135. How many weeks should she wait before she can afford the outfit, if sales tax is $8^3/_4$ percent?

16. Mr. Miller invested $1500 4 years ago in a fund and hasn't made any withdrawals. If the average annual rate of interest over the 4 years was 12 percent, how much money is currently in the fund?

17. A certain credit card charges 18 percent annual interest on the unpaid monthly balance. Mike purchased a computer for $1300, which already includes sales tax, using this credit card. If he pays $200 every month, how long will it take Mike to pay off his debt? What was his total cost for the computer?

18. Gary found that his attic needed more insulation to prevent loss of heat. He has to be careful about how much weight he adds to the attic. Insulation comes in rolls that weigh 7 lb 4 oz. Each roll will insulate 40 ft². If his attic is rectangular and measures 15 ft by 18 ft, how many rolls of insulation will be needed and how much weight, to the nearest pound, will this insulation add to the attic?

19. Mrs. Alexander, who lives in New York, would like to purchase three pieces of furniture made in England that she saw in a catalog. The cost of shipping by air across the Atlantic Ocean is $365 per 100 lb. In addition, there is a 3 percent international tariff. If the three pieces weigh 16, 32, and 51 kg, how much money would it cost to import these pieces of furniture from London?

20. European Airways allows each passenger to check two suitcases that measure 50 cm by 30 cm by 72 cm. The airline assumes that luggage weighs 75 kg per cubic meter. What is the maximum weight, in pounds, of luggage that the plane would be carrying if there are 200 passengers on board?

Solutions to Additional Problems

1. The total perimeter of the garden is $10^1/_2$ ft $+ 10^1/_2$ ft $+ 8$ ft $= 29$ feet. In order to calculate the cost, Gayle needs to convert this to

meters. Using the conversion scheme of creating fractions, we have

$$\frac{29\,\text{ft}}{1} \times \frac{12\,\text{in}}{1\,\text{ft}} \times \frac{2.54\,\text{cm}}{1\,\text{in}} \times \frac{1\,\text{m}}{100\,\text{cm}} = 8.84\,\text{m}$$

The total cost is $8.84\,\text{m} \times \$4.25/\text{m} = \$37.57$

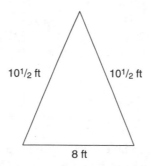

2. The distance from the 8-ft side to the opposite vertex is the height of the triangle (draw this line in the diagram from the previous problem and label it 6 ft). The base is the $10\frac{1}{2}$-ft side. Using the formula for the area of a triangle from the reference table, $A = \frac{1}{2} \times \text{base} \times \text{height}$, we have $A = 0.5 \times 10.5 \times 6 = 31.5\,\text{ft}^2$.

3. The perimeter of the frame is given as 50 in. (Note how the perimeter is disguised as the amount of wood used to make the frame.) If the longer side of the picture is 11 in and there is a 2-in border on either side, then the longer side of the frame has to be 15 in. The two longer sides of the frame, therefore, use 30 in of the wood and the formula for the perimeter of a rectangle, $P = 2 \times \text{length} + 2 \times \text{width}$, tells us that each of the shorter sides has to be half of the remaining 20 inches. The shorter side of the frame is 10 in.

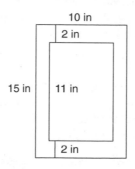

4. The gallon of paint would cover 400 ft². Assuming that we would have a rectangular area painted yellow at the top of the wall, we could use the 25 feet to represent the base of the rectangle. The formula for the area of a rectangle, A = base × height, tells us that $400 = 25$ × height or height $= 400 \div 25 = 8$ ft. This is exactly one-half of the wall.

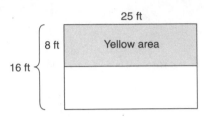

5. Since the pool has to be 6 ft from the property line, the diameter of the pool can be no more than $26 - 12 = 14$ ft. Therefore, the radius of the largest possible circular area is 7 ft. Using the formula for the area of a circle, $A = \pi \times \text{radius}^2$, we have $A = 3.14 \times 7^2 = 153.86$ ft².

6. To solve this problem, we need to calculate the area of the entire rectangular yard and subtract the area of the circle. Since the price of the sod is given in dollars per square foot, we need to convert the measurements of the yard to feet. These conversions are:

$$\frac{12\,m}{1} \times \frac{100\,cm}{1\,m} \times \frac{1\,in}{2.54\,cm} \times \frac{1\,ft}{12\,in} = 39.4\,ft = 39\,ft$$

$$\frac{14\,m}{1} \times \frac{100\,cm}{1\,m} \times \frac{1\,in}{2.54\,cm} \times \frac{1\,ft}{12\,in} = 45.9\,ft = 46\,ft$$

Using the formula for the area of a rectangle, A = length × width, the area of the entire yard is $39 \times 46 = 1794$ ft². Using the formula for the area of a circle, $A = \pi \times \text{radius}^2$, the circular design has an area of $A = 3.14 \times 10^2 = 314$ ft². Therefore, Melissa needs to buy $1794 - 314 = 1480$ ft² of sod. This will cost her $1480\,\text{ft}^2 \times \$6/\text{ft}^2 = \$8880$.

7. We know the volume of the tank to be 5520 L. Since we are being asked to find a height, we have to be working in measurements of length. We therefore need to convert liters into cubic meters in order to work with the same units. From the reference table, we

20

see that 1 m³ is equal to 1 kiloliter (kL) or 1000 L. Therefore, the tank will hold $5520 \div 1000 = 5.52$ m³. We need to calculate the volume of the tank. The formula for the volume of a rectangular three-dimensional shape is V = area of the base $\times$ height. Using this, we have 5.52 m³ $= 2$ m² $\times$ height (in meters). Therefore, the height is $5.52 \div 2 = 2.76$ m. Notice how, in this last equation, we have to have equal units on each side, m³ $=$ m³. This is a good way to be sure that you are applying formulas correctly.

8. We need to know exactly how much ice cream is used in the ice-cream cone. Since *gallons* is mentioned in the problem, we know that we will be calculating volume. Using the formulas for the volumes of a cone and of a sphere from the formula reference table, we have to identify the radius and the height of the cone and the radius of the scoop. In both cases, the diameter is 2 in; therefore, the radius of each is 1 in. We apply the formulas Volume of cone $= \frac{1}{3} \times 3.14 \times 1^2 \times 6 = 6.28$ in³ and Volume of scoop $= \frac{4}{3} \times 3.14 \times 1^3 = 4.18$ in³, and the total volume is 10.46 in³. However, we need to convert this to gallons since the cost is given in dollars per gallon. The reference table tells us that there are 277.42 in³ in 1 gal. Therefore, the number of gallons is $10.46 \div 277.42 = 0.0377$ or 0.04 gal. The ice-cream parlor owner's cost in making the cone is 0.04 gal $\times$ \$30/gallon $=$ \$1.20. The owner is making a profit of \$1.95 $-$ \$1.20 $=$ \$0.75 on the cone.

9. The total time taken for the flight is 4 hours and 45 minutes. To use this in the formula for *distance traveled*, we have to convert this to a decimal number in hours. We can use (45 minutes/1) $\times$ (1 hour/ 60 minutes) $= 0.75$ hour. Our time is, therefore, 4.75 hours. Applying the formula, we have Distance traveled $= 320$ mph $\times$ 4.75 hours $= 1520$ mi.

10. Using the speed limit of 65 mph in the formula for *distance traveled*, we have 680 mi $= 65$ mph $\times$ the number of hours taken. Therefore, the number of hours is $680 \div 65 = 10.46$, or approximately $10\frac{1}{2}$ hours. It would have taken Jennifer 1 hour more. If needed, you could also calculate her rate using the formula $680 =$ average speed $\times$ 9.5 or average speed $= 680 \div 9.5 = 71.57$ or 72 mph, which means that she was traveling faster than the speed limit allowed.

11. The formula for distance traveled is applied to determine the record speed in meters per second: 100 m $=$ average speed $\times$ 9

seconds or speed $= 100 \div 9 = 11.11$ m/second. This has to be converted to miles per hour, and we can use the "fraction" scheme for doing this:

$$\frac{11.11\,\text{m}}{1\,\text{second}} \times \frac{60\,\text{seconds}}{1\,\text{minute}} \times \frac{60\,\text{minutes}}{1\,\text{hour}} \times \frac{1\,\text{km}}{1000\,\text{m}} \times \frac{1\,\text{mi}}{1.61\,\text{km}}$$

$$= 24.84\,\frac{\text{mi}}{\text{hour}}$$

To the nearest tenth, a runner would have to run faster than 24.9 mph to beat the record.

12. To find the time taken, we need to convert the Celsius temperature to Fahrenheit. We can use the formula $F = \frac{9}{5} \times C + 32$ to get $F = 1.8 \times 42 + 32 = 107.6°F$ or $108°F$ (where $F =$ degrees Fahrenheit and $C =$ degrees Celsius, as in Example 12). Therefore, the temperature has to rise $108 - 68 = 40°F$. At $2°$ per minute, it would take 20 minutes for the compound to boil.

13. By applying the formula $C = \frac{5}{9} \times (F - 32)$ to 75°F and to 85°F, we have a lower temperature of $\frac{5}{9} \times (75 - 32) = 23.9°C$ and a higher temperature of $\frac{5}{9} \times (85 - 32) = 29.4°C$.

14. The discount has to be applied first. That is, one of the pairs of jeans will sell for the full price of $24.99 while the other will sell for one-half of that, or $12.50. Therefore, the 6 percent sales tax has to be applied to the sum of the prices, $37.49. The total cost of the purchase is found from the formula Total cost = price + tax × price, and we compute $37.49 + .06 \times $37.49 = 39.74.

15. In this problem it is important to remember that the $90 that Lindsey has to spend must include the $8\frac{3}{4}$ percent sales tax. We can find the different costs of the outfit as the weeks pass by making successive computations and keeping a list. Using the formula Discounted price = original price − discount × original price, the price after the first week would be $135 − $135 × 0.10 = $121.50. After the second week, it would be $121.50 − $121.50 × 0.10 = $109.35. After the third week, the price would be $109.35 − $109.35 × 0.10 = $98.42. After the fourth week, the price would be below $90; that is, $98.42 − $98.42 × 0.10 = $88.58. We have to compute the price with tax to see if Lindsey's $90 would be enough. The total price would be $88.58 + 0.0875 \times $88.58 = $96.33, which is more than she could spend. The price after the fifth week would be

$88.58 − $88.58 × .10 = $79.72. The total price with tax would be $79.72 + 0.0875 × $79.72 = $86.70. Therefore, Lindsey would have to wait 5 weeks and hope that the outfit hasn't been sold.

16. Each year there is a new amount that is earning interest. (This is usually called *compounding interest*.) Using the formulas Interest earned = principal × rate of interest × time invested and Current amount = principal + interest earned, we can determine how much Mr. Miller has at the end of each year.

At end of year	Interest earned	New amount
1	$1500 × 0.12 = $180	$1500 + $180 = $1680
2	$1680 × 0.12 = $201.60	$1680 + $201.60 = $1881.60
3	$1881.60 × 0.12 = $225.79	$1881.60 + $225.79 = $2107.39
4	$2107.39 × 0.12 = $252.89	$2107.39 + $252.89 = $2360.28

A table like this is a good tool for keeping track of calculations and gives you a way to look back on your work in an organized fashion.

17. This problem is similar to one asking for a total amount earned with interest. That is, a new amount is owed each month after adding on interest on the unpaid balance and $200 deduction. A table is useful to follow the month-by-month situation. Since the interest is added monthly, we need to use 18 percent ÷ 12 = 1.5 percent to calculate the added interest each month. Before beginning the calculations, an easy estimate should be made. Since $1300 ÷ $200 = 6.5, we would expect that with interest adding to the amount, it should take 7 or perhaps 8 months to pay off the debt.

At end of month	Interest added	Total debt	New balance
1	$1300 × 0.015 = $19.50	$1300 + $19.50 = $1319.50	$1319.50 − $200 = $1119.50
2	$1119.50 × 0.015 = $16.79	$1119.50 + $16.79 = $1136.29	$1136.29 − $200 = $936.29
3	$936.29 × 0.015 = $14.04	$936.29 + $14.04 = $950.33	$950.33 − $200 = $750.33
4	$750.33 × 0.015 = $11.25	$750.33 + $11.25 = $761.58	$761.58 − $200 = $561.58
5	$561.58 × 0.015 = $8.42	$561.58 + $8.42 = $570.00	$570.00 − $200 = $370.00
6	$370.00 × 0.015 = $5.55	$370.00 + $5.55 = $375.55	$375.55 − $200 = $175.55
7	$175.55 × 0.015 = $2.63	$175.55 + $2.63 = $178.18	Pay $178.18 to complete debt

We see that our estimate was a good one. The total cost of the computer is found by adding the interest amounts to the original price of $1300; that is, $1300 + $78.18 = $1378.18.

18. First we need to determine how many rolls of insulation are needed. The area of the attic is 15 ft × 18 ft = 270 ft². Therefore, we need 270 ÷ 40 = 6.75 or 7 rolls. For the second part of the question, we can use the exact number 6.75 since we do not have to use the remaining portion of the roll. First, we need to convert 7 lb 4 oz into a decimal number in pounds. Since there are 16 oz in one pound, 4 oz is ¼ or 0.25 lb. Therefore, each roll weighs 7.25 lb. The weight is therefore 6.75 rolls ×7.25 lb/roll = 48.9375 or 49 lb.

19. The answer is found by breaking the problem down into smaller parts.

Step 1 Find the total weight in kilograms and convert it to pounds. The total weight is 16 + 32 + 51 = 99 kg. Converting to pounds, we have 99 kg × 2.2 lb/kg = 217.8 lb.

Step 2 Find the cost per pound. The shipping rate is given with respect to 100 lb. Dividing by 100 gives the cost per pound, $3.65 per pound.

Step 3 Find the shipping charge. The formula from the reference table

$$\text{Total dollar amount of item} = \text{quantity of item} \times \text{dollar amount of one item}$$

is applied to get 217.8 lb × $3.65/lb = $794.97.

Step 4 Compute the tax to add, $794.97 × 0.03 = $23.85.

Step 5 The total charge is $794.97 + $23.85 = $818.82.

20. **Step 1** The dimensions of the suitcase should be converted to meters in order to use the weight rate. Dividing each measurement by 100 gives us 0.5, 0.3, and 0.72 m.

Step 2 The suitcase has a volume of 0.5 m × 0.3 m × 0.72 m = 0.108 m³.

Step 3 The weight of one suitcase would be .108 m³ × 75 kg/m³ = 8.1 kg.

Step 4 There could be up to 200 × 2 = 400 suitcases on the plane for a total weight of 400 × 8.1 = 3240 kg.

Step 5 Multiplying by 2.2 would give us the number of pounds; That is, 3240 kg ×2.2 lb/kg = 7128 lb.

24

Table of Common Measurements and Conversions

Length or Distance in Linear Units

Straight-line distance between 2 points: 1 unit

English system		Metric system		Conversions between systems	
1 inch (in)	Basic unit	1 millimeter (mm)	1/1000 m	1 in	2.54 cm
1 foot (ft)	12 in	1 centimeter (cm)	1/100 m	39 in	1 m
1 yard (yd)	36 in	1 decimeter (dm)	1/10 m	1 mi	1.61 km
1 mile (mi)	5280 ft	1 meter (m)	Basic unit		
		1 kilometer (km)	1000 m		

Area or Space in Square Units

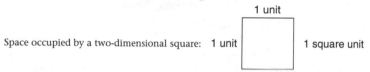

Space occupied by a two-dimensional square: 1 unit   1 square unit

English system		Metric system		Conversions between systems	
1 square inch (in^2)	Basic unit	1 square meter (m^2)	Basic unit	10.8 ft^2	1 m^2
1 square foot (ft^2)	144 in^2	1 hectare (ha)	10,000 m^2	2.47 acres	1 ha
1 square yard (yd^2)	9 ft^2				
1 acre	43,560 ft^2				

Volume or Capacity in Cubic Units

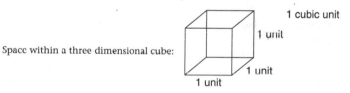

1 cubic unit

Space within a three dimensional cube:

English system		Metric system		Conversions between systems	
1 ounce (oz)*	Basic unit	1 milliliter (mL)	1 cm^3 or cc	1.06 quarts	1 liter
1 cup	4 oz*	1 liter (L)	1000 mL or 1 dm^3		
1 pint (pt)	2 cups = 8 oz				
1 quart (qt)	2 pt = 32 oz	1 kiloliter (kL)	1000 L or 1 m^3		
1 gallon (gal)	4 qt or 277.42 in^3				

Weight or Mass

English system		Metric system		Conversions between systems	
1 ounce (oz)[†]	Basic unit	1 gram (g)	Basic unit	2.2 lb	1 kg
1 pound (lb)	16 oz[†]	1 kilogram (kg)[‡]	1000 g		
1 ton	2000 lb	1 metric ton (t)	1000 kg		

*These are referred to as *fluid* ounces.
[†]These are referred to as *dry* ounces.
[‡]One kilogram is equal to the weight of one liter of water.

Table of Common Formulas Seen in Word Problems

Measuring Length

Perimeter of a rectangle $= 2 \times$ length $+ 2 \times$ width
or $2 \times$ base $+ 2 \times$ height:

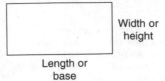

Width or height

Length or base

Circumference of a circle $= 2 \times \pi \times$ radius (π is approximately 3.14)
Distance traveled $=$ average rate $\times$ time traveled

Measuring Area

Area of a rectangle $=$ length $\times$ width or base $\times$ height

Area of a square $=$ side^2:

 Side

Area of a parallelogram $=$ base $\times$ height:

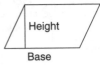

Height

Base

Area of a triangle $= \frac{1}{2} \times$ base $\times$ height:

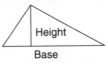

Height

Base

Area of an equilateral triangle $= \sqrt{3}/4 \times$ side^2:

Side

Area of a trapezoid $=$ average of the bases
$\times$ height or $\frac{1}{2} \times$ (lower base $+$ upper base) $\times$ height:

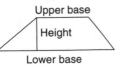

Upper base

Height

Lower base

Area of a circle $= \pi \times$ radius^2 (π is approximately 3.14):

Radius

Continue

Surface area of a three-dimensional sphere = $4 \times \pi \times \text{radius}^2$:

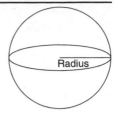

Measuring Volume

Volume of a rectangular solid = length × width × height or area of the base × height:

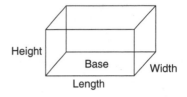

Volume of a cube = side^3:

Volume of a sphere = $\frac{4}{3} \times \text{radius}^3$

Volume of a circular cylinder = $\pi \times \text{radius}^2 \times \text{height}$:

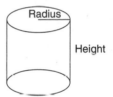

Volume of a circular cone = $\frac{1}{3} \times \pi \times \text{radius}^2 \times \text{height}$:

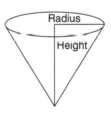

Money-Related Formulas

Total dollar amount of item = quantity of item × dollar amount of one item
Total cost = price + tax rate × price
Discounted price = original price − discount × original price
Interest earned = principal × rate of interest × time invested
Current amount = principal + interest earned

Chapter 2

Using Algebraic Equations to Solve Problems

In the previous chapter, you studied problems that were solved by making basic arithmetic computations. In this chapter, you will learn how to apply algebra to solve problems. In most cases, this will require *modeling* the situation with an equation in which what you wish to find will be expressed by a letter representing an unknown quantity. We call this the *variable* of the equation. The equations will come from applying basic formulas and well-known concepts. (You can review basic tools of algebra by visiting the Appendix in the back of the book.)

It is important to carefully identify the *unknown variable* in the problem and determine the relationship between it and the other information in the problem. It is also important that your answer really does solve the problem! There are many places to make mistakes in these problems, such as creating the wrong equation, solving the equation incorrectly, or making an arithmetic mistake.

Problems Involving Numbers

The simplest of all word problems are those that involve finding numbers. Most of these problems will tell you how two or more numbers are related. The key will be to identify all the relationships between the numbers that are mentioned in the problem.

Example I

The larger of two numbers is equal to 14 more than 3 times the smaller number. Find the smaller number if their difference is 38.

Solution I

The first step of any word problem is to find a way to use a variable to represent the unknowns in the problem. Since the problem specifically tells us how the larger and smaller numbers are related, we can use one variable. It is also a good idea to select a letter for the variable that reminds us of what we're looking for. The letter n is always a good choice for a number.

The first sentence of this problem tells us that if we know the smaller number, we can compute the larger one. It makes sense, therefore, to use the variable to represent the smaller one. We should also start our work with a statement of how we are assigning variables. This is often called the "Let" statement.

Let the smaller number $= n$

We are now ready to start translating the first sentence into an algebraic equation. A good idea is to first rewrite the problem leaving most of the words and using mathematical symbols for adding, subtracting, multiplying, dividing, and equality when it is easily recognized. This would give us

Larger number $= 14$ more than $3 \times$ smaller number

Now we have to take a look at the phrase on the right. We must realize that to have *more than* something means to *add* to that something.

Larger number $= 3 \times$ smaller number $+ 14$

We can now substitute our variable expressions for the smaller and larger parts. Remember to drop the multiplication symbol when multiplying by a variable.

Larger number $= 3n + 14$

Both numbers in the problem are now represented by variables, and we are ready to form the equation that we will use to solve the problem. This will come from the other relationship given in the problem, "their difference is 38." Again we should first translate the operation symbols and write the remaining words. We should realize that finding a difference means to subtract, and if the difference is positive, we are taking the smaller from the larger. We get

$$\text{Larger number} - \text{smaller number} = 38$$

Substituting our variables gives us

$$3n + 14 - n = 38$$

Simplifying and solving for n, we get

$$2n + 14 = 38$$
$$2n = 24$$
$$n = 12$$

Therefore, the smaller number is 12, and the larger number is $3 \times 12 + 14 = 36 + 14 = 50$. To check, we need only verify that their difference is 38. Clearly, $50 - 12 = 38$. ✓

Sometimes one of the relationships regarding the numbers is given indirectly; that is, it might be hidden in a key phrase.

Example 2

Separate 70 into two parts so that 5 times the smaller part is equal to 14 more than twice the larger part.

Solution 2

The problem is really asking for two different numbers whose sum is 70. We can assign a variable, n, to the smaller part. The larger part must be what is left when n is taken away from 70, $70 - n$. The "Let" statement is

$$\text{Let } n = \text{smaller part}$$
$$70 - n = \text{larger part}$$

Translating only the operations gives us

$5 \times$ smaller part $= 14$ more than twice the larger part

Now we have to take a look at the phrases on the right: specifically, "more than" and "twice." We must realize that to have *more than* something means to *add* to that something.

$5 \times$ smaller part $=$ twice the larger part $+ 14$

Now we use the fact that to have twice something means to multiply that something by 2.

$5 \times$ smaller part $= 2 \times$ larger part $+ 14$

Now we are ready to substitute our variable expressions for the smaller and larger parts. Remember to drop the multiplication symbol and to *use parentheses* when we need to multiply by an expression with more than one term.

$$5n = 2(70 - n) + 14$$

Solving in the usual way, we get

$$5n = 140 - 2n + 14$$

Add $2n$ to both sides to get

$$7n = 154$$
$$n = 22$$

The smaller part is 22, and the larger part is $70 - 22 = 48$.

The check is easy. Simply follow the numbers and words of the problem, 5 times the smaller is $5 \times 22 = 110$; 14 more than twice the larger is $2 \times 48 + 14 = 96 + 14 = 110.\checkmark$

Alternative Solution 2

Another way to solve the problem is to use two different variables to represent both the parts of 70. When we use two

32

variables, we need two equations. (See the Appendix for a review of how to work with equations with two variables.)

$$\text{Let } x = \text{the smaller number}$$

$$\text{Let } y = \text{the larger number}$$

The first equation comes from the fact that the sum of the numbers is 70.

$$x + y = 70$$

The second equation comes from the way these parts are related. The translation should still be performed as outlined above, but now we use the variables x and y to substitute for the smaller part and for the larger part.

$$5x = 2y + 14$$

We work with this pair by solving the first equation for y

$$y = 70 - x$$

and substituting in the second equation

$$5x = 2(70 - x) + 14$$

Note that we have the exact same equation to solve (even though the variable here is x instead of n) and we should arrive at the solution, $x = 22$ and $y = 48$.

The next example shows you another way to express one of the relationships indirectly.

Example 3

In this example, 5 less than 6 times the largest of three consecutive positive integers is equal to the square of the smaller minus 2 times the middle integer. Find all three integers.

Solution 3

It is clear that we are dealing with three different numbers. The key phrase in this problem is *consecutive positive integers*. To understand this, let's look at each word. The numbers have to be integers. This is the set of negative and positive whole numbers and zero $\{\ldots, -3, -2, -1, 0, 1, 2, 3, \ldots\}$; that is, we won't accept any answers that are fractions, decimals, or roots. Since the numbers are positive, we are looking only for numbers from the set $\{1, 2, 3 \ldots\}$. The most important word in the phrase is *consecutive*. This means that the three numbers are all one apart. We can then represent our numbers with the same variable, n.

Let $n =$ the smallest integer

$n + 1 =$ the middle integer

$n + 2 =$ the largest integer

Now we can use our translation scheme. First only the symbols

5 less than 6 × largest = square of the smallest

− 2 times the middle

Less than something means to take away from that something

6 × largest − 5 = square of the smallest − 2 × middle

Square of means to raise the number to the second power

6 × largest − 5 = smallest^2 − 2 × middle

Now we can substitute our variables and remove the multiplication symbols.

$$6(n + 2) - 5 = n^2 - 2(n + 1)$$

This is a *quadratic equation* since we have an equation with a power of 2. (See the Appendix for a review of solving quadratic equations.) It must be simplified and have 0 on one side in order to solve it by factoring.

$$6n + 12 - 5 = n^2 - 2n - 2$$

To both sides of the equation, we subtract n^2, add $2n$, and add 2, to arrive at

$$-n^2 + 8n + 9 = 0$$

We can now multiply every term in the equation by -1 to get

$$n^2 - 8n - 9 = 0$$

We can now factor and solve for n.

$$(n-9)(n+1) = 0$$

Therefore

$$n = 9 \quad \text{or} \quad n = -1$$

Remember, we are looking for *positive* integers, so we can *reject* -1 as an answer. Since n was the smallest, the three integers are 9, 10, and 11.

To check, we simply use the numbers and follow the words of the problem: 5 less than 6 times the largest is $6 \times 11 - 5 = 66 - 5 = 61$.

The square of the smallest minus 2 times the middle is $9^2 - 2 \times 10 = 81 - 20 = 61.\checkmark$ (Not all consecutive integer problems will give rise to quadratic equations as you will see in the supplementary exercises.)

Problems Involving Age

Problems that involve the ages of two or more people are very similar to number problems. (After all, an age is a number.) Since we need to have two relationships, these problems may cause you to consider ages at the present time and at another time in the past or future.

Example 4

Sandy is 4 times as old as Robby is now. In 10 years, Sandy will be 6 years older than Robby. How old are they now?

Solution 4

Since we are told how to find Sandy's age if we know Robby's age, we should use a variable for Robby's age and determine the symbolic representation of Sandy's age with the same variable. Following the procedure outlined in the previous problems, we have

$$\text{Let Robby's age} = x$$
$$\text{Sandy's age} = 4 \times \text{Robby's age}$$
$$\text{Sandy's age} = 4x$$

"In 10 years" means that we will add 10 to each of their ages. We ought to state what their ages will be.

$$\text{In 10 years, Robby's age will be } x + 10$$
$$\text{and Sandy's age will be } 4x + 10$$

We should now translate the second sentence using these future ages and substitute the future variables

$$\text{Sandy's age} = \text{Robby's age} + 6$$
$$4x + 10 = x + 10 + 6$$

Simplifying and solving for x, we get

$$4x + 10 = x + 16$$
$$3x = 6$$
$$x = 2$$

Therefore, Robby's current age is 2 and Sandy's current age is 8.

To check, we should simply follow the words of the problem. Sandy's current age is 4 times Robby's age or $8 = 4 \times 2$.✓

In 10 years, Sandy will be 18 and Robby will be 12. Clearly, Sandy will be 6 years older than Robby.✓

Example 5

Jessica is 27 years old and Melissa is 21 years old. How many years ago was Jessica twice as old as Melissa?

36

In this problem, the variable is the number of years in the past, and we have to subtract this amount from their current ages.

Let x = the number of years in the past

Therefore, x years ago, Jessica was $27 - x$ years old and Melissa was $21 - x$ years old. The relationship x years ago was

$$\text{Jessica's age} = 2 \times \text{Melissa's age}$$

Substituting the past ages, we get the equation for the problem.

$$27 - x = 2(21 - x)$$

Simplifying and solving for x, we get

$$27 - x = 42 - 2x$$
$$x = 15 \text{ years ago}$$

To check, we have to compute the girls' ages 15 years ago. Jessica was $27 - 15 = 12$ years old and Melissa was $21 - 15 = 6$ years old. Clearly, Jessica was twice as old as Melissa. ✓

In Chap. 1 we solved problems that make use of many of the basic formulas that arise in mathematics. Those problems merely required that we identify the appropriate formula, substitute known values, and solve for the missing value. Now we will examine situations where those formulas are used as the basis for the equations needed to solve more complex problems. In each problem, we will see that the key to the problem is finding out which of the quantities mentioned are meant to be equal or which quantities are added to get a total amount.

Problems Involving Motion

In problems that involve motion, there are usually one or two objects moving along a straight line at a constant rate (speed) for a period of time. The two objects can be moving toward each other, moving away from each other, or moving in the same direction as in a race. In any situation, there is a total distance traveled by the objects. The governing relationship for these problems is

$$\text{Distance} = \text{rate} \times \text{time}$$

Example 6

Two cars are traveling on the same highway toward each other from towns that are 150 mi apart. One car is traveling at 60 mph, while the other is traveling at 50 mph. If they started at the same time, how much time, to the nearest minute, will have passed at the moment when they pass each other?

Solution 6

In order to use the formula, we have to decide what is given and what we have to find by examining each of the parts. Drawing a picture can help. The car traveling at 60 mph will cover more of the distance than the car traveling at 50 mph.

The picture shows that the total distance traveled is a known amount and that it must be equal to the distances traveled by each car. This is the basis of the model for this problem.

$$\text{Total distance} = \text{distance traveled by faster car}$$
$$+ \text{distance traveled by slower car}$$

We know the rates of each car, and the unknown is the time. Since the cars started at the same time and pass each other at the same time, the time they've both traveled is the same. Therefore, we can use one variable for this time, t.

As before, it is a good idea to state the use of the variable before any work is begun.

$$\text{Let } t = \text{time traveled by both cars}$$

Using the relationship Distance = rate × time for each car, our model becomes

$$150 = 60t + 50t$$
$$150 = 110t$$
$$t = \frac{150}{110}$$
$$t = 1.363636\ldots$$

Remember that this answer must be in hours since the rates were given in miles per *hour*. The problem asks for the answer to be given to the nearest minute. The decimal part of the hour can be converted to minutes by multiplying .37 × 60 = 22.2 minutes and, when rounded, the answer is 1 hour and 22 minutes.

To check the answer, we need to compute the distances traveled by each car. To be accurate, we will use the actual fraction $15/11$ for the time. The faster car traveled $15/11$ × 60 = $900/11$, and the slower car traveled $15/11$ × 50 = $750/11$. By adding these fractions, we find that together they traveled $1650/11$ = 150 mi.✓

Example 7

Allison and Rachel were competing in a race. Allison had a 10-minute head start and was running at 3 km per hour. Rachel ran at 5 km per hour. In how many minutes did Rachel catch up with Allison?

Solution 7

In this problem, the key factor is that at the moment that Rachel catches up to Allison, they will have traveled the same distance. The model for the problem is

Distance traveled by Allison = distance traveled by Rachel

Since we are asked to find Rachel's time, we will represent it by t. Note that the units of t must be hours, since the rates are given in hours. Allison ran 10 minutes more, which will

39

be represented by $t + \frac{1}{6}$, since 10 minutes is one-sixth of an hour.

$$\text{Let } t = \text{Rachel's time}$$
$$t + \tfrac{1}{6} = \text{Allison's time}$$

Using Distance = rate × time, we have

$$3\,(t + \tfrac{1}{6}) = 5t$$
$$3t + \tfrac{1}{2} = 5t$$
$$\tfrac{1}{2} = 2t$$
$$t = \tfrac{1}{2} \div 2 = \tfrac{1}{4}\,\text{hours} \quad \text{or} \quad t = 15\,\text{minutes}$$

To check, we compute the distance traveled by each girl using $t = \frac{1}{4}$ for Rachel and $t = \frac{1}{4} + \frac{1}{6} = \frac{5}{12}$ for Allison. Rachel's distance was $5 \times \frac{1}{4} = 1.25$ km, and Allison traveled $3 \times \frac{5}{12} = 1.25$ km as well.

Note that we could have converted the rates to kilometers per minute by dividing each rate by 60. This would allow t to represent minutes, and we could use 10 minutes in the equation instead of $\frac{1}{6}$ hour.

$$\left(\frac{3}{60}\right)(t + 10) = \left(\frac{5}{60}\right)t$$
$$\frac{3t}{60} + \frac{30}{60} = \frac{5t}{60}$$
$$3t + 30 = 5t$$
$$30 = 2t$$
$$15 = t$$

The relationship used in these problems can be rewritten to express the rate as a quotient:

$$\text{Rate} = \frac{\text{distance traveled}}{\text{time}}$$

This form of the relationship may be necessary to use in problems when the rates aren't explicitly given.

Problems Involving Work

There are other situations which follow the same principle using a rate. For example, when someone works at a job, we can use

Number of jobs completed = rate of work × time

or, equivalently

$$\text{Rate} = \frac{\text{number of jobs completed}}{\text{time}}$$

Example 8

If Joe can paint five rooms in 8 hours and Samantha can paint three rooms in 6 hours, how long will it take them to paint seven rooms working together?

Solution 8

Using the rate formula, we see that Joe's rate of work is $5/8$ rooms per hour and Samantha's rate of work is $1/2$ room per hour.

This problem is actually similar to Example 6, in that they both work for the same amount of time and together they complete the given total amount.

Let t = the time to complete the job together

The model is

Total number of rooms painted

= rooms painted by Joe + rooms painted by Samantha

$7 = \frac{5}{8}t + \frac{1}{2}t = \frac{5}{8}t + \frac{4}{8}t$

$7 = \frac{9}{8}t$

$56 = 9t$

$t = \frac{56}{9} = 6\frac{2}{9}$ hours

Example 9

It takes Suzanne 10 hours longer than Laurie to take their store's inventory. If it takes $3\frac{3}{4}$ hours for them to do the job together, how long would it take Suzanne to do it alone?

Solution 9

Since Laurie takes less time to complete the job, we will let her time be represented by t. If we can find this time, then the answer to the problem will be $t + 10$.

The number of jobs considered in this problem is 1. Therefore, Laurie's rate is $1/t$ and Suzanne's rate is $1/(t + 10)$. The model for the problem is similar to Example 8, where the total number of jobs completed is the sum of the amount of work done by each girl in the 2 hours.

1 job = amount done by Suzanne + amount done by Allison

Using $\frac{15}{4}$ for $3\frac{3}{4}$, the amount done by Suzanne is $\frac{1}{t+10} \times \frac{15}{4}$, and the amount done by Laurie is $\frac{1}{t} \times \frac{15}{4}$. Substituting into the preceding model, we have

$$1 = \frac{1}{t+10} \times \frac{15}{4} + \frac{1}{t} \times \frac{15}{4} = \frac{15}{4}\left(\frac{1}{t+10} + \frac{1}{t}\right)$$
$$\frac{1}{\frac{15}{4}} = \frac{1}{t+10} + \frac{1}{t}$$

(Note that this line tells us that the *combined rate is equal to the sum of the individual rates*. If you remember this fact, you can use it as a model when the problem involves only completing 1 job.) Simplify the fraction on the left to get the fractional equation

$$\frac{4}{15} = \frac{1}{t+10} + \frac{1}{t}$$

Multiplying both sides of the equation by the least common

42

denominator, $15t(t+10)$, we get

$$4t(t+10) = 15t + 15(t+10)$$

Simplifying, we get a quadratic equation $\qquad 4t^2 + 40t = 30t + 150$

Subtracting $30t$ and 150 from both sides gives us $\qquad 4t^2 + 10t - 150 = 0$

Dividing all terms by 2 simplifies the equation to $\qquad 2t^2 + 5t - 75 = 0$

We can now factor and solve $\qquad (2t+15)(t-5) = 0$

$$2t + 15 = 0 \text{ or } t - 5 = 0$$
$$t = -7.5 \text{ or } t = 5$$

Since the problem requires physical time, we *reject* the negative answer and we see that $t = 5$ is Laurie's time. Therefore, Suzanne can complete the job in 15 hours working alone.

Problems Involving Mixing Quantities

A situation similar to those in the problems above is finding out what occurs when we mix different types of objects. In these problems there is also a rate involved, but it is not a rate involving time, as we will see.

Example 10

At the chocolate factory, John's job is to experiment with mixing different amounts of milk and chocolate to find different flavors. If he mixes 2 quarts (qt) of type A, which contains 15 percent chocolate, with 1 qt of type B, which contains 10 percent chocolate, he creates type C. What percent of type C is chocolate?

Solution 10

In this situation, the rate is the percentage of chocolate. For each type, we have the relationship

Amount of chocolate in the type

$=$ percent of chocolate $\times$ amount of the type

43

where the percentage is taking the form of the rate as we saw in previous problems.

In type A, we have $0.15 \times 2 = 0.3$ qt of chocolate

In type B, we have $0.10 \times 1 = 0.1$ qt of chocolate

Together, in the mixture type C, we have 0.4 qt of chocolate. Since the mixture has exactly 3 qt of liquid, we calculate the percentage that is chocolate. To do this, we rewrite the relationship (amount of chocolate)/(amount of type) = percentage of chocolate. Since there are $0.1 + 0.3 = 0.4$ qt, the answer is easily found to be $0.4/3 = 0.1333\ldots$ or 13.3%.

Another type of mixture problem involves the price of a mixture of two items where each sells for different individual prices. The underlying formula is also a kind of rate relationship since price is usually given as dollars per pound, cents per pound, or dollars per kilogram. For instance

$$\frac{\text{Cost}}{\text{amount}} = \text{price} \quad \text{or} \quad \text{cost} = \text{price} \times \text{amount}$$

Example 11

Nancy's Nut Stand sells cashews for $5/lb and honey-roasted peanuts for $2/lb pound. How many pounds of each type of nut should Nancy use to make 8 lb of a mixture that will sell for $3/lb?

Solution 11

The problem asks us to find the amount of both kinds of nuts. We can solve the problem using only one variable since we know that the total amount must be 8 lb.

Let p = number of pounds of cashews

$8 - p$ = number of pounds of honey-roasted peanuts

The model for the problem is that the total cost for the mixture should be the same as the sum of the costs for each individual kind of nut:

Cost of mixture = cost of cashews

+ cost of honey-roasted peanuts

44

and using the relationship Cost = price × amount, we have

$$(3)(8) = 5p + 2(8 - p)$$
$$24 = 5p + 16 - 2p$$
$$8 = 3p$$
$$p = \frac{8}{3} \quad \text{or} \quad 2\frac{2}{3}$$

The mixture should have $2\frac{2}{3}$ lb of cashews and $8 - 2\frac{2}{3} = 5\frac{1}{3}$ lb of honey-roasted peanuts.

Alternative Solution 11

Another way to solve the problem is to use two variables and get two equations to work with. (See the Appendix for a review of how to work with equations with two variables.)

Let x = amount of cashews in mixture

Let y = amount of honey-roasted nuts in mixture

The first equation is the total amount of the mixture

$$x + y = 8$$

The second equation comes from the cost relationship Cost = price × amount, and the model Cost of cashews + cost of honey-roasted peanuts = cost of mixture.

$$5x + 2y = (3)(8) = 24$$

The two equations together form a simultaneous system of equations and can be solved easily by the following process.

Multiply the first equation by 2 and subtract it from the first

$$5x + 2y = 24$$
$$2x + 2y = 16$$
$$3x = 8$$
$$x = \frac{8}{3} = 2\frac{2}{3}$$

45

Use this value of x in the first equation and solve for y:

$$2\,^2/_3 + y = 8$$
$$y = 5\,^1/_3$$

It's okay if you are not familiar with working with two equations. As we saw, all of the problems can be solved with one variable and one equation. However, you will be a better problem solver by having several ways to solve a problem in your arsenal.

Problems Involving Money

One other kind of problem that is easily solved by an algebraic equation involves combinations of money of different denominations. These are problems similar to the previous ones because the amount of money that a coin is worth is a sort of rate. For example, a quarter is worth 25 cents per coin and a dime is worth 10 cents per coin. The amount of money we have is found by the relationship

Amount of money = value of coin × number of coins

Here, the *value of the coin* is taking the form of the rate.

Example 12

Joey has $2.35 in nickels, dimes, and quarters. If he has two less quarters than dimes and one more nickel than dimes, how much of each coin does he have?

Solution 12

The key to the problem is the way in which the numbers of quarters and nickels are related to the number of dimes. We can find the amounts of quarters and nickels if we know the amount of dimes. Therefore, we will assign the variable to this amount.

Let x = number of dimes

Since the number of quarters is two less than this amount, we have

$$x - 2 = \text{number of quarters}$$

Since the number of nickels is one more than the number of dimes, we have

$$x + 1 = \text{number of nickels}$$

The governing relationship is

$$\text{Total amount of money} = \text{amount in quarters}$$
$$+ \text{amount in dimes} + \text{amount in nickels}$$

Using the Amount = rate of coin × number of coins, we have

$$2.35 = .25(x - 2) + 0.10x + 0.05(x + 1)$$

Since we are given the total amount in dollars, we need to use decimals to express the rate of each coin as a decimal fraction of a dollar. If we multiply every term by 100, we can clear the decimals

$$235 = 25(x - 2) + 10x + 5(x + 1)$$

Simplifying and solving for x, we have

$$235 = 25x - 50 + 10x + 5x + 5$$
$$235 = 40x - 45$$
$$280 = 40x$$
$$7 = x$$

Therefore, there are seven dimes, five quarters, and eight nickels. This is easily checked: 70¢ + $1.25 + 40¢ is indeed $2.35.✓

Summarizing what we have done with all our word problems, we saw that we need to follow some basic steps:

Step 1

Read the problem carefully and identify the unknowns.

Step 2

Identify the formula or key concept that is needed in the problem and rewrite the problem, keeping most of the words but using basic arithmetic symbols for operations and equality.

Step 3

Assign a variable to one of the unknowns with a "Let" statement and determine the symbolic representations of all the other knowns with this variable using the relationships in the problem.

Step 4

Substitute the representations of the unknowns into the mathematical statement you created in step 2.

Step 5

Use algebra to solve the equation(s) and determine the value of the unknowns you represented by a variable.

Step 6

Check your answer to see if it really does solve the problem and satisfy all the necessary conditions.

It is worth remembering that in many problems, there are at least two relationships among the unknowns. If only two unknown quantities are asked for, the mathematical model of the problem is sometimes more easily seen when using two variables and creating a system of two equations to solve.

Additional Problems

1. The sum of two numbers is 62. The larger number is as much greater than 34 as the smaller number is greater than 10. What are the two numbers?
2. Separate the number 17 into two parts so that the sum of their squares is equal to the square of one more than the larger part.

3. The difference between two numbers is 12. If 2 is added to 7 times the smaller, the result is the same as when 2 is subtracted from 3 times the larger. Find the numbers.

4. The sum of three consecutive integers is 11 greater than twice the largest of the three integers. What are they?

5. Find three consecutive odd integers whose sum is 75.

6. Find four consecutive even integers such that the sum of the squares of the two largest numbers is 34 more than the sum of the other two.

7. A father is now 5 times as old as his son. In 15 years he will be only twice as old as his son. How old are they now?

8. Sarah's mom is 42 years old, and her aunt is 48 years old. How many years ago was Sarah's aunt 3 times as old as her mom?

9. A cyclist traveling 20 mph leaves ½ hour ahead of a car traveling at 50 mph. If they go in the same direction, how long will it take for the car to pass the cyclist?

10. In the early days of building railroads in the United States, the first transcontinental railroad was started from two points at the ends of the 2000-mi distance over which tracks had to be laid. Two teams at different ends of the country started at the same time. The team in the east moved forward at 2 mi per day over mostly flat land, and the team from the west moved at 1.5 mi per day over mountainous terrain. How long did it take for the teams to meet and the last spike to be nailed in, and how many miles of track did each team lay?

11. On an international flight across the Atlantic Ocean, a certain type of plane travels at a speed that is 220 km/hour more than twice the speed of another. The faster plane can fly 7000 km in the same time that the slower plane can fly 2700 km. At which speeds do these planes fly in miles per hour?

12. Mr. Begly's office has an old copying machine and a new one. The older machine takes 6 hours to make all copies of the financial report, while the new machine takes only 4 hours. If both machines are used, how long will it take for the job to be completed?

13. The candystand at the multiplex has 100 lb of a popular candy selling at $1.20/lb. The manager notices a different candy worth $2.00/lb that isn't selling well. He decides to form a mixture of both types of candy to help clear his inventory of the more expensive type. How many pounds of the expensive candy should he mix with the 100 lb in order to produce a mixture worth $1.75/lb?

14. At her specialty coffee store, Melissa has two kinds of coffee she'd like to mix to make a blend. One sells for $1.60/lb and the other, $2.10/lb. How many pounds of each kind must she use to make

75 lb of mixed coffee that she can sell for $1.90/lb and make the same profit?

15. Joey has $5.05 in 34 coins on his dresser. If the coins are quarters and dimes, how many of each kind are there?

16. The triplets John, Jen, and Joey gave their father all their coins toward a present for their mother. John had only quarters to give. Jen had only dimes and gave 4 times as many coins as John gave. Joey had only nickels and gave seven more than twice the number of coins that Jen gave. If the total amount of the contribution was $22.40, how much did each child contribute?

17. Mrs. Stone likes to diversify her investments. She invested part of $5000 in one fund that had a return of 9% interest and the rest into a different fund that earned 11%. If her total annual income from these investments was $487, how much does she invest at each rate?

18. Three people each received part of $139 so that the first received twice what the second received and the third's portion exceeded the sum of the other two by one dollar. How much money did each person receive?

19. A farmhand was hired for a 60-day period and would live on the ranch. For each day that he worked he would receive $45, but for each day that he did not work he would pay $10 for his room and board. At the end of the 60 days, he received $2260. How many days did he work?

Solutions to Additional Problems

1. Using one variable, we have the following. Let x = the smaller number; therefore, $62 - x$ is the larger number. In Prob. 1, the expression "as much greater than" indicates that we want the difference between the numbers.

$$\text{The larger number} - 34 = \text{the smaller number} - 10$$
$$(62 - x) - 34 = x - 10$$
$$28 - x = x - 10$$
$$38 = 2x$$
$$x = 19$$

The smaller number is 19, and the larger number is $62 - 19 = 43$.

Check: $43 - 34 = 9$ and $19 - 10 = 9$.✓ An alternative solution using two variables is as follows. Let x = smaller number and let y = larger number. The equations would be $x + y = 62$ and $y - 34 = x - 10$. The second equation could be rewritten as $-x + y = 24$. Adding the first equation to this new form of the second equation gives us $2y = 86$ and $y = 43$. Substituting into the first equation, we have $x + 43 = 62$ and $x = 19$.

2. Both "parts" of the number 17 can be represented using one variable. We should use the single variable to represent the larger part, since another piece of the puzzle relates to it. Let x = the larger part and $17 - x$ = smaller part and $x + 1$ = one more than the larger part. Recognizing the *sum of their squares* to mean *larger part^2 + smaller part^2*, we have the quadratic equation

$$x^2 + (17 - x)^2 = (x + 1)^2$$

Squaring the binomials, we have

$$x^2 + 289 - 34x + x^2 = x^2 + 2x + 1$$

Combining terms gives us

$$2x^2 - 34x + 289 = x^2 + 2x + 1$$

Subtracting terms on the right from both sides of the equation produces

$$x^2 - 36x + 288 = 0$$

We can factor this and solve

$$(x - 12)(x - 24) = 0$$
$$x - 12 = 0 \quad \text{or} \quad x - 24 = 0$$
$$x = 12 \quad \text{or} \quad x = 24$$

Since the larger part of 17 must be less than 17, we reject the answer of 24. The larger part is 12 and the smaller part is 5.

Check: $12 + 5 = 17$. $12^2 + 5^2 = 144 + 25 = 169$ and $(12 + 1)^2 = 13^2 = 169$.✓

3. The problem is best understood and modeled when two variables are used for larger and smaller numbers. The model is

$$\text{Larger number} - \text{smaller number} = 12$$

$$7 \times \text{smaller} + 2 = 3 \times \text{larger} - 2$$

Let x = larger number and let y = smaller number. The two equations are $x - y = 12$ and $7y + 2 = 3x - 2$. We can rewrite the second equation as $-3x + 7y = -4$, and we will be able to eliminate one of the variables by adding equations if we multiply every term in the first equation by 7: $7x - 7y = 84$. Adding the equations gives us $4x = 80$ and $x = 20$. Using the original first equation to find y, we have $20 - y = 12$ and $y = 8$. The numbers are 20 and 8.

Check: The difference of the numbers is $20 - 8 = 12$. The other relationships are

$$7 \times 8 + 2 = 56 + 2 = 58$$

$$3 \times 20 - 2 = 60 - 2 = 58\checkmark$$

4. *Consecutive* integers imply that the three numbers are one apart. Therefore

$$\text{Let } x = \text{first (smallest) number}$$
$$x + 1 = \text{second (middle) number}$$
$$x + 2 = \text{third (largest) number}$$

The model is

$$\text{First number} + \text{second number} + \text{third number}$$

$$= 2 \times \text{third number} + 11$$

(Be sure not to confuse the phrase "is 11 greater than" with "11 is greater than," which translates into an inequality "11 >"!) Then $x + (x + 1) + (x + 2) = 2(x + 2) + 11$. This linear equation can be solved in the usual way. (At this point, see if you can identify what steps are being taken.)

$$3x + 3 = 2x + 4 + 11$$
$$3x + 3 = 2x + 15$$
$$x = 12$$

The numbers are 12, 13, and 14.

Check: 12, 13, and 14 are *consecutive* integers. $12 + 13 + 14 = 39$ and $2 \times 14 + 11 = 28 + 11 = 39.\checkmark$

5. By definition, an *odd number* is an integer (fractions are not considered to be odd or even). To be consecutive odd numbers the numbers must differ by 2. Therefore

$$\text{Let } x = \text{first (smallest) odd number}$$
$$x + 2 = \text{second (middle) odd number}$$
$$x + 4 = \text{third (largest) odd number}$$

The model is simply

$$\text{First odd number} + \text{second odd number}$$
$$+ \text{third odd number} = 75$$

The equation is therefore

$$x + (x + 2) + (x + 4) = 75$$
$$3x + 6 = 75$$
$$3x = 69$$
$$x = 23$$

The three consecutive odd numbers are 23, 25, and 27.

Check: $23 + 25 + 27 = 75.\checkmark$

6. Consecutive even integers must be 2 apart. Therefore,

$$\text{Let } x = \text{first (smallest) even integer}$$
$$x + 2 = \text{second even integer}$$
$$x + 4 = \text{third even integer}$$
$$x + 6 = \text{fourth (largest) even integer}$$

The model for the problem is

$$(\text{Third integer})^2 + (\text{fourth integer})^2 = \text{first integer}$$
$$+ \text{second integer} + 36$$

53

The equation is therefore

$$(x + 4)^2 + (x + 6)^2 = x + (x + 2) + 34$$
$$x^2 + 8x + 16 + x^2 + 12x + 36 = 2x + 36$$
$$2x^2 + 20x + 52 = 2x + 36$$
$$2x^2 + 18x + 16 = 0$$
$$x^2 + 9x + 8 = 0$$
$$(x + 1)(x + 8) = 0$$
$$x + 1 = 0 \quad \text{or} \quad x + 8 = 0$$
$$x = -1 \quad \text{or} \quad x = -8$$

Since we want even integers, we reject $x = -1$ and our smallest even integer is -8. The four even integers are -8, -6, -4, and -2.

Check: $(-4)^2 + (-2)^2 = 16 + 4 = 20$ and $-8 + -6 + 34 = 20.\checkmark$

Don't be disturbed by having negative numbers as answers. There is no reason to assume that all number problems involve only positive integers!

7. The two relationships are clearly stated in the problem. One relationship can serve as the basis for assigning a variable. Since the father's age is given in terms of the son's age

$$\text{Let } x = \text{son's age now}$$
$$5x = \text{father's age now}$$

The second relationship provides us with the model

$$\text{Father's age now} + 15 = 2 \times (\text{son's age now} + 15)$$
$$5x + 15 = 2(x + 15)$$
$$5x + 15 = 2x + 30$$
$$3x = 15$$
$$x = 5$$

The son is now 5 and the father is $5 \times 5 = 25$.

Check: In 15 years the son will be 20 and the father will be 40. $40 = 2 \times 20.\checkmark$

54

8. The unknown in this problem is the number of years in the past from now. The model for the problem is

> Sarah's aunt's age in the past
> $= 3 \times$ Sarah's mom's age in the past

Let $x =$ the number of years from then until now. Sarah's aunt's age was $48 - x$. Sarah's mom's age was $42 - x$. The equation is therefore

$$48 - x = 3(42 - x)$$
$$48 - x = 126 - 3x$$

To both sides we add $3x$ and subtract 48 to get the simple equation $2x = 78$. Dividing both sides by 2 gives us $x = 39$. Therefore, 39 years ago, Sarah's aunt was 3 times as old as Sarah's mom.

Check: 39 years ago, Sarah's aunt was $48 - 39 = 9$ and Sarah's mom was $42 - 39 = 3$. $9 = 3 \times 3$.✓

9. The "hidden" concept here is that at the moment when the car passes the cyclist, they have traveled the same distance. The model for the problem will be

> Distance traveled by car $=$ distance traveled by cyclist

For each distance, we make use of the formula Distance $=$ rate $\times$ time. Since time is the unknown and the cyclist travels 0.5 hour longer than the car, we have

> Let $t =$ time traveled by car
> $t + 0.5 =$ time traveled by cyclist

The equation we have from the model is

$$20 \times (t + 0.5) = 50t$$
$$20t + 10 = 50t$$
$$10 = 30t$$
$$t = \frac{1}{3} \text{ hour or 20 minutes}$$

Check: The cyclist travels for $\frac{1}{3} + \frac{1}{2} = \frac{5}{6}$ hours and covers a distance of $20 \times \frac{5}{6} = \frac{100}{6}$ or $\frac{50}{6}$ mi. The car travels for $\frac{1}{3}$ hour and covers a distance of $50 \times \frac{1}{3} = \frac{50}{3}$ mi. ✓

10. This is really a "motion" problem when thinking of the teams as very slow moving vehicles. The model for the problem is based on the fact that the total distance is equal to the sum of the distances covered by both teams or

> Distance covered by team from east
>
> + distance covered by team from west = 2000

The "hidden" fact here is that when two objects are moving along the same path toward each other, then at the point when they meet, they have spent the same amount of time traveling! Since the rate is given in miles per *day*, we let t = the number of days spent by both teams. Then, using Distance = rate × time for each team's distance, we have

$$2\,\text{mi/day} \times t + 1.5\,\text{mi/day} \times t = 2000$$

$$2t + 1.5t = 2000$$

$$3.5t = 2000$$

$$t = 2000 \div 3.5 = 571.43\,\text{days}$$

or approximately $571.43/365 = 1.56$ or $1\frac{1}{2}$ years. The team from the east laid approximately $2 \times 571.43 = 1142.8$ or 1143 mi of track. The team from the west laid approximately $1.5 \times 571.43 = 857.1$ or 857 miles of track.

Check: Since the problem called for the amount of track laid, the only necessary check is that $1143\,\text{mi} + 857\,\text{mi} = 2000\,\text{mi}$.

11. Notice that the problem specifies that we give our answer in miles per hour. This adds only one more step in the solution as we can find the rates in kilometers per hour and convert to miles per hour at the end. Working in the metric system does not change the way in which we solve the problem.

There are clearly two relationships given in the problem. The first is straightforward and allows us to use one variable to represent both rates.

$$\text{Let } r = \text{rate of slower plane}$$

$$2r + 220 = \text{rate of faster plane}$$

The problem gives us the distances for the same amount of time traveled by both planes. From Distance = rate × time, we have the alternative form of Time = distance/rate, and we can derive the model

$$\frac{\text{Distance of fast plane}}{\text{Rate of fast plane}} = \frac{\text{distance of slow plane}}{\text{rate of slow plane}}$$

and the equation

$$\frac{7000}{2r + 220} = \frac{2700}{r}$$

Cross-multiplying, we have

$$7000r = 2700(2r + 220)$$

Simplifying and solving, we obtain

$$7000r = 5400r + 594,000$$

$$1600r = 594,000$$

$$r - 371.25$$

The slower plane is traveling at 371.25 km/hour and the faster plane is traveling at 962.50 km/hour. Multiplying each of these rates by the conversion factor of 0.62 mi/km, we have that the slower plane is traveling at approximately 230 mph and the faster plane can travel at 597 mph.

Check: For the check, we will use the metric answers. The slow plane travels 2700 km in $2700 \div 371.25 = 7.2727 \ldots$ hours. In this time, the fast plane travels $962.5 \times 7.2727 \ldots = 7000$ km.✓

12. The model here states that if the machines are working together for the same amount of time, each will do a fraction of the job.

Therefore

$$1 \text{ job} = \text{fraction done by old machine}$$
$$+ \text{fraction done by new machine}$$

The rate of each machine can be found from

$$\text{Rate} = \frac{\text{number of jobs completed}}{\text{time}}$$

The rate of the older machine is $1/6$ job/hour, and the rate of the new machine is $1/4$ job/hour. Let $t =$ the number of hours the machines work together.

Using the Number of jobs completed = rate of work × time for each machine, we have

$$1 = \tfrac{1}{6}t + \tfrac{1}{4}t$$
$$1 = \tfrac{5}{12}t$$
$$t = \tfrac{12}{5} \text{ or } 2.4 \text{ hours}$$

Check: In 2.4 hours the older machine has done $2.4/6 = 0.4$ of the job and the newer machine has done $2.4/4 = 0.6$ of the job; thus $0.4 + 0.6 = 1$ job.✓

13. The underlying concept here is that the amount of money earned from selling the mixture at $1.75 per pound should equal the total of having sold each of the two candies at their normal price or

Dollar amount from mixture

= dollar amount from cheaper candy

+ dollar amount from expensive candy

We use the formula Dollar amount = price × amount sold. Let $x =$ the number of pounds of expensive candy and we have that $x + 100$ be the amount of the mixture. The equation from the model is $1.75(x + 100) = 1.20 \times 100 + 2.00x$. Simplifying the equation and multiplying every term by 100 gives us

$$175x + 17,500 = 12,000 + 200x$$
$$5500 = 25x$$
$$x = 220 \text{ lb}$$

Check: When all candy is sold, the mixture would bring in $1.75 × 320 = $560; specifically, 100 lb of the cheaper candy would bring in $1.20 × 100 = $120, and 220 lb of the expensive candy would bring in $2.00 × 220 = $440. $120 + $440 = $560.✓

14. The phrase "make the same profit" indicates that Melissa is seeking to receive the same amount of money from the blend as she would if she sold the same amounts that are in the mixture separately. The model is therefore

Dollar amount received from blend

= dollar amount from cheaper coffee

+ dollar amount from the expensive coffee

The two amounts of coffee can be represented by the same variable easily.

Let x = number of pounds of cheaper coffee in blend

$75 - x$ = number of pounds of expensive coffee in blend

Each amount is found by the formula Dollar amount = price × amount used. Therefore, our equation is

$$1.90 \times 75 = 1.60x + 2.10(75 - x)$$
$$142.50 = 1.60x + 157.5 - 2.10x$$
$$-15 = -0.5x$$
$$x - 30$$

Melissa needs to use 30 lb of the cheaper coffee and 45 lb of the more expensive coffee.

Check: The amount from the blend is $142.50, specifically, $1.60 × 30 = $48.00 and $2.10 × 45 = $94.50. $48.00 + $94.50 = 142.50.✓

15. The two relationships in this problem are

Total number of coins = 34

Value of the coins = $5.05

Let $d =$ the number of dimes and let $q =$ the number of quarters. The first relationship gives us the equation $d + q = 34$. The second relationship uses the following concepts.

Amount of money $=$ value of coin $\times$ number of coins

Total amount of money $=$ amount in quarters

$+$ amount in dimes

The second equation is therefore $0.10d + 0.25q = 5.05$. To get both equations into a form in which we can add or subtract them to eliminate one of the variables, we multipy the first equation by 25 and the second equation by 100 to get

$$25d + 25q = 850$$
$$10d + 25q = 505$$

and we subtract to get $15d = 345$. Therefore, $d = 23$, and from the original first equation

$$23 + q = 34 \qquad \text{or} \qquad q = 11$$

Check: 23 dimes $+$ 11 quarters $= 34$ coins and $\$.10 \times 23$ $+ \$.25 \times 11 = \$2.30 + \$2.75 = \$5.05.\checkmark$ (Try solving the problem, using d for the number of dimes and $34 - d$ for the number of quarters.)

16. The model for the problem is given by

Amount from John's quarters $+$ amount from Jen's dimes

$+$ amount from Joey's nickels $= \$22.40$

Each amount is calculated by Amount of money $=$ value of coin $\times$ number of coins. We should try to express all three numbers of coins with the same variable instead of using three different variables. To do this, we need to study the given relationships carefully. Let $x =$ number of coins (quarters) John gave. Jen gave 4 times as many coins as John, so

$$4x = \text{number of coins (dimes) Jen gave}$$

Joey gave 2 times the number of coins Jen gave + 7, so

$$8x + 7 = \text{number of coins (nickels) Joey gave}$$

Using these representations in the model, we have

$$0.25x + 0.10 \times 4x + 0.05(8x + 7) = 22.40$$

Simplifying and multiplying every term by 100, we have

$$25x + 40x + 40x + 35 = 2240$$
$$105x + 35 = 2240$$
$$105x = 2205$$
$$x = 21$$

Using this in our "Let" statements, we deduce that John gave 21 quarters, Jen gave $4 \times 21 = 84$ dimes, and Joey gave $8 \times 21 + 7 = 175$ nickels.

Check: 84 is $4 \times 21 \checkmark$ and 175 is $2 \times 84 + 7$.$\checkmark$ Also

$$\$0.25 \times 21 + \$0.10 \times 84 \mid \$0.05 \times 175$$
$$= \$5.25 + \$8.40 + \$8.75 = \$22.40 \checkmark$$

17. This problem is similar to those we've looked at in that the underlying model is

Total interest earned = interest earned from 9% fund
+ interest earned from 11% fund

We use the formula

Interest earned = principal × rate of interest
× time invested

and note that the time over which this interest was earned was 1 year. The unknown in the problem is the amount invested in each fund. Since the total amount is $5000, we let $x =$ the amount invested at 9% and $5000 - x =$ the amount invested at 11 percent. The model gives us the equation

$$487 = 0.09x + 0.11(5000 - x)$$
$$48,700 = 9x + 55,000 - 11x$$
$$48,700 = 55,000 - 2x$$
$$2x = \$6300$$
$$x = \$3150$$

Mrs. Stone invested $3150 in the 9% fund and the balance, $5000 - \$3150 = \$1850$ in the 11 percent fund.

Check: The interest from the 9 percent fund was $0.09 \times \$3150 = \$283.50$, and the interest from the 11 percent fund was $0.11 \times \$1850 = \$203.50$; then $\$283.50 + \$203.50 = \$487.00.\checkmark$

18. It would become very complicated to try to use three variables to represent the different amounts that each person received. This forces us to try to use only one variable to model the problem. The situation is clearest when we use the second person's amount as the single variable since the first is easily determined when we know this amount. Let $x =$ amount received by second person. Clearly, $2x =$ amount received by first person.

The third person must have received the remaining portion of the $139 or $139 - (x + 2x)$.

$$139 - 3x = \text{amount received by third person}$$

Note that the phrase "exceeds by some amount" can be translated as "is that amount more than." The model is, therefore

$$\text{Third person's portion} = \text{first person's portion}$$
$$+ \text{second person's portion} + \$1$$
$$139 - 3x = 2x + x + 1$$
$$139 - 3x = 3x + 1$$
$$138 = 6x$$
$$x = 23$$

The second person received $23, the first person received $46, and the third person received $139 − $69 = $70.

Check: $23 + $46 + $70 = $139; then $23 + $46 + $1 = $70.✓

19. This problem doesn't belong to any category that was mentioned in this chapter, but it is still basically a "rate × amount" problem and is easily solved with some algebra. The rates are $45 earned per day and $10 paid per day. Having two variables makes the modeling easy. (See the Appendix for a review of how to work with equations having two variables.)

Let x = number of days that the farmhand worked

Let y = number of days that the farmhand did not work

We know that the total number of days is 60. Therefore, the first equation is

$$x + y = 60$$

The basic concept is that

Amount earned from working

− amount paid for room and board = amount received

The second equation is, therefore, $45x − 10y = 2260$. Multiplying every term in the first equation by 10 and adding this to the second equation gives

$$10x + 10y = 600$$
$$45x − 10y = 2260$$

Adding the two equations gives us $55x = 2860$, and we find $x = 52$. The farmhand worked for 52 days and did not work for 8 days.

Check: 52 + 8 = 60; then 52 × $45 −8 × $10 = $2340 − $80 = $2260.✓

Chapter 3

Word Problems Involving Ratio, Proportion, and Percentage

Many situations presented in word problems involve comparison of the sizes of groups and objects. The answer sought may not be the number of objects, but the fraction that indicates how much of the larger group is the smaller or how two groups are related to the whole. These fractions are referred to as *ratios*.

Sometimes when ratios are given, they are given by the statement $a:b$. This is equivalent to the fraction a/b. The fraction allows us to use arithmetic in the usual way to solve problems.

Example I

In a math class, there are 32 students. If the ratio of boys to girls is 3:5, how many boys and how many girls are there in the class?

Solution I

One way to understand this problem is to restate the ratio in the following way. For every three boys there are five girls. With this we can create a visual representation of the situation using B for a boy and G for a girl. Writing groups of BBBGGGGG and keeping track of how many we have in a cumulative way gives us:

Groups	Running count
BBBGGGGG	8
BBBGGGGG	16
BBBGGGGG	24
BBBGGGGG	32

Adding the Bs and Gs tells us that there are 12 boys and 20 girls. The check for this answer is to create the fraction of boys/girls and see if it is equal to $3/5$.

Check: $12/20 = 0.6 = 3/5.\checkmark$

Alternative Solution 1

The method described above was quick, but it would be tedious if the group were much larger than 32. The problem really involves finding the number of groups and applying a rate. The ratio 3 boys:5 girls can be translated into two rates:3 boys/group and 5 girls/group. The model for the situation is

Class size = number of boys + number of girls

where each number is found by

Rate of boys or girls per group × the number of groups

(In Chap. 2, there are many problems using similar models.)
    If we think of the number of groups as the unknown, we can solve the problem algebraically by assigning a variable to this and modeling the situation with an equation.

Let x = the number of groups

Number of boys = $3x$

Number of girls = $5x$

Using these quantities in the model, we have the equation

$$3x + 5x = 32$$
$$8x = 32$$
$$x = 4$$

There are four groups, each consisting of 3 boys and 5 girls. Therefore, there are $3 \times 4 = 12$ boys and $5 \times 4 = 20$ girls.

Example 2

An advertisement for toothpaste claims that 4 out of 5 dentists recommend a certain brand. A consumer group wanted to check the accuracy of this claim and surveyed 180 dentists. How many dentists would they expect to not recommend this brand, if the advertisement's claim is accurate?

Solution 2

The phrase "4 out of 5" indicates that the ratio of those who recommend to those who do not recommend is 4:1. Using the picture scheme R for recommend and D for doesn't recommend, we would have to list and count groups of RRRRD. Rather than listing groups, we easily realize that there has to be $180/5 = 36$ groups and, therefore, there would be 36 Ds if the claim were accurate.

The more structured algebraic approach would be

$$\text{Let } x = \text{the number}$$
$$\text{of groups}$$

$$\text{Number of those who recommend} = 4x$$

$$\text{Number of those who do not recommend} = x$$

$$4x + x = 180$$

$$5x = 180$$

$$x = 36$$

There would be $4 \times 36 = 144$ dentists who recommend the brand and $1 \times 36 = 36$ who do not.

Ratios can appear in many of the types of problems we saw in the previous chapter.

Number Problems

Example 3

Two positive numbers are in the ratio of 5:8. Find the numbers, assuming that the square of the smaller is 15 less than 10 times the larger.

Solution 3

If we think of multiples of 5 and 8 as "groups," we can use a solution similar to that for Probs. 1 and 2; specifically, we can list pairs of multiples and check the conditions for each pair.

Multiple	Smaller	Larger	Smaller^2	10 × larger−15
1	5	8	25	$80 - 15 = 65$
2	10	16	100	$160 - 15 = 145$
3	15	24	225	$240 - 15 = 225\checkmark$

As before, this is an easy solution because the actual numbers came up early in the list. If the actual numbers were larger, listing them in this fashion would be tedious.

The algebraic approach would be

$$\text{Let } x = \text{"multiple" needed}$$

$$5x = \text{smaller number}$$

$$8x = \text{larger number}$$

The equation would come from the model

$$\text{Smaller}^2 = 10 \times \text{larger} - 15$$

and we would have the quadratic equation:

$$(5x)^2 = 10(8x) - 15$$

$$25x^2 = 80x - 15$$

$$25x^2 - 80x + 15 = 0$$

Dividing every term in the equation by 5 gives us the simpler quadratic

$$5x^2 - 16x + 3 = 0$$

$$(5x - 1)(x - 3) = 0$$

$$5x - 1 = 0 \text{ or } x - 3 = 0$$

$$x = \tfrac{1}{5} \quad \text{or} \quad x = 3$$

Since x is a multiple, it must be a whole number. Therefore, we reject the answer of $\frac{1}{5}$. Our numbers would be the third multiples of 5 and 8 or 15 and 24.

Check: $15^2 = 225$; then $10 \times 24 - 15 = 240 - 15 = 225.\checkmark$

Age Problems

Example 4

Jason is Chris' older cousin. The ratio of their ages is 9:5. In 5 years, Chris' age will be 4 more than half of Jason's age. How old are the boys now?

Solution 4

The ratio indicates that the current ages of the boys are multiples of 5 and 9. Making a table as follows will lead to the answer quickly.

Now		In 5 years		$\frac{1}{2}$ Jason's age	Chris' age $-\frac{1}{2}$ of Jason's age
Chris	Jason	Chris	Jason		
5	9	11	15	$7\frac{1}{2}$	$-3\frac{1}{2}$
10	18	15	23	$11\frac{1}{2}$	$+3\frac{1}{2}$
15	27	20	32	16	$+4$ $\checkmark$

Alternative Solution 4

Using algebra

$$\text{Let } x = \text{the multiple}$$

$$\text{Chris' current age} = 5x$$

$$\text{Jason's current age} = 9x$$

The model is

$$\text{Chris' age} + 5 = \frac{1}{2} \text{ (Jason's age} + 5) + 4$$

Substitution gives

$$5x + 5 = \frac{1}{2}(9x + 5) + 4$$

$$5x + 5 = 4.5x + 2.5 + 4$$

$$5x + 5 = 4.5x + 6.5$$
$$0.5x = 1.5$$
$$x = 3$$

Chris' age is $5 \times 3 = 15$, and Jason's age is $9 \times 3 = 18$.

Check: In 5 years their ages would be 20 and 32. Specifically, $20 = \frac{1}{2} \times 32 + 4 = 16 + 4 = 20.\checkmark$

Perimeter

Example 5

The perimeter of a triangular plot of land is 273 m. A surveyor is making a map of the land and notes that the ratio of the sides is 2:9:10. What is the actual length of each side of the plot?

Solution 5

In solving this problem, the use of a table would be a lengthy task. An algebraic approach gets to the heart of the problem quickly. (Make sure to draw a diagram that reflects the ratios to help visualize the problem accurately.)

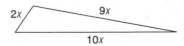

Let x be the multiple.
The sides of the triangular plot are $2x$, $9x$, and $10x$.
The model is

$$\text{Sum of all sides} = 273$$

We have the equation

$$2x + 9x + 10x = 273$$
$$21x = 273$$
$$x = 13$$

70

The sides are therefore $2 \times 13 = 26$ m, $9 \times 13 = 117$ m, and $10 \times 13 = 130$ m.

Check: $26 + 117 + 130 = 273.\checkmark$

Mixture

Example 6

Carole's candystand sells a popular mixture of two of her candies which sell for \$2.70/lb and \$1.95/lb. The ratio of the expensive candy to the cheaper candy in the mixture is 3:5. If Carole sold her entire supply of the mixed candy during the coming month, she would have revenue of \$928.20, which would be the same as if she sold each candy separately. How many pounds of candy does she have in supply, and at what price does she sell it for?

Solution 6

This seems like a more complicated problem than the previous ones in that more information is given. The key to the problem is that the revenue would be the same as if she sold each type of candy separately. Sometimes when a problem seems complicated, we should try to use a simple number and work with the information in the problem to help get a better understanding of what is occurring. For example, since the ratio of the candies in the mixture is 3:5, let's assume that 8 lb was sold. In other words, we might ask, "What would be the revenue if 3 lb of the expensive candy and 5 lb of the cheaper candy were sold?"

Clearly, the answer to this question is $3 \times \$2.70 + 5 \times \$1.95 = \$17.85$. The price of the mixture would be $\$17.85 \div 8$ lb, giving us \$2.23/lb (rounded).

We easily see that 8 lb is not the answer we are looking for, but it helped us model the problem. The model is

Number of pounds of expensive candy $\times$ \$2.70 + number of pounds of cheaper candy $\times$ \$1.95 = \$928.20

71

Using algebra:

Let $x =$ the multiple

Number of pounds of expensive candy $= 3x$

Number of pounds of cheaper candy $= 5x$

Total number of pounds sold $= 8x$

The model gives us the equation

$$3x \times 2.70 + 5x \times 1.95 = 928.20$$
$$17.85x = 928.20$$
$$x = 52$$

Therefore, $8 \times 52 = 416$ lb of the mixture is in supply and sells for $928.20 \div 416 = \$2.23$/lb. (Note that this is the same price as when 8 lb is sold. This would be the price for any amount sold, since we are maintaining the same ratio in all quantities of the mixture.)

Check: The mixture contains $3 \times 52 = 156$ lb of the expensive candy, which would have a revenue of $156 \times \$2.70 = \$421.20$, and $5 \times 52 = 260$ lb of the cheaper candy, which would have a revenue of $260 \times \$1.95 = \$507.00$. The total revenue is $\$421.20 + \$507.00 = \$928.20.\checkmark$

Proportion

Word problems sometimes involve comparing two situations that are numerically or geometrically similar. This often requires a scaling of the quantities in a *proportional* manner. For example, a numerically similar situation could be working with a recipe that lists quantities required to make four servings needs to be scaled to make more servings when more than four are required. A geometrically similar situation could be working with two rectangles that are of different sizes but of the same length:width ratio.

When dealing with such situations, we create a mathematical statement indicating that the ratios are the same. This is called a *proportion* and would appear in ratio form as $a{:}b = c{:}d$ or in fraction form as $a/b = c/d$.

72

From the ratio form, we call a and d the *extremes* while b and c are referred to as the *means*. From the fraction form, we see that if we multiply both sides of the equation by bd, we arrive at $ad = bc$. This is sometimes referred to as *cross-multiplying in order to solve a proportion*. The more mathematical phrase is: *In a proportion, the product of the means is equal to the product of the extremes*. This will be the governing principle in solving the following problems.

Measurement of Quantities

Example 7

Among the ingredients in a recipe for a sweet dessert are $2\frac{1}{2}$ cups of regular sugar and $\frac{1}{4}$ cup of brown sugar. The directions indicate that this will serve 8 people. If there will be 12 people to serve, how much of each type of sugar will be needed?

Solution 7

The underlying assumption in the problem is that the recipe needs to be scaled proportionally when seeking to make more servings. The first step is to decide which two of the three quantities given in the original recipe are necessary for the ratios that will appear in the proportion: $2\frac{1}{2}$, $\frac{1}{4}$, or 8. It is helpful in solving problems of this kind to rephrase the question so that the problem is better understood: "What happens when the number of people changes to 12?" This indicates that our ratios and proportion must involve 8 and 12.

This also helps us see that we will need to solve two different proportions: one for regular (refined white) sugar and the other for brown sugar. The model for the problem is, therefore

(*a*) The amount of regular sugar for 8 people *is proportional to* the amount of regular sugar for 12 people.

(*b*) The amount of brown sugar for 8 people *is proportional to* the amount of brown sugar for 12 people.

Using (*a*), we can assign a variable and proceed algebraically. Let x be the amount of regular sugar for 12 people. The model

73

gives us the equation

$$\frac{2\frac{1}{2}}{8} = \frac{x}{12}$$

Cross-multiplying, we have $8x = 30$ and $x = 30/8 = 3\frac{3}{4}$ cups of regular sugar. Using (b), we can solve in a similar fashion. Let y be the amount of brown sugar for 12 people

$$\frac{\frac{1}{4}}{8} = \frac{y}{12}$$

Cross-multiplying, we have $8x = 3$ and $y = \frac{3}{8}$ cup of brown sugar.

Check: One way to check the problem is to see if the ratios of the original amounts to the new amounts is the same as 12:8 or $\frac{3}{2}$. For regular sugar, $3\frac{3}{4} \div 2\frac{1}{2} = \frac{15}{4} \div \frac{5}{2} = \frac{15}{4} \times \frac{2}{5} = \frac{30}{20} = \frac{3}{2}.\checkmark$

For brown sugar, $\frac{3}{8} \div \frac{1}{4} = \frac{3}{8} \times \frac{4}{1} = \frac{12}{8} = \frac{3}{2}.\checkmark$

Measurement of Similar Figures

Example 8

On a blueprint for a new office building, the rectangular conference room measures 4.5 in by 12 in. The shorter wall of the actual room measures 15 ft. How much carpeting will be needed to cover the floor of the actual room?

Solution 8

The underlying assumption is that a blueprint is a *scaled* drawing of the room; that is, the sides of the room on the blueprint are proportional to the actual sizes or the ratios of the corresponding sides are equal. The model for the problem is (length on blueprint)/(actual length) = (width on blueprint)/(actual width).

Once we find the actual width, we need to compute the area of the floor of the room. Letting w be the actual width, we have $4.5/15 = 12/w$. Cross-multiplying gives us $4.5w = 180$. Therefore, $w = 180/4.5 = 40$ ft. The area is $15 \times 40 = 600$ ft^2.

In Chap. 1, we were concerned about how to keep track of units of measure. Note that in computation of w, we actually have the following calculation, including units: $w = (12 \text{ in} \times 15 \text{ ft})/4.5 \text{ in} = 40 \text{ ft}$, which shows that we have made the correct calculation as it gives us our answer in feet when we cancel units.

We see that something interesting happened when we compare the areas of the blueprint drawing to the actual drawing. The area of the room on the blueprint is $4.5 \text{ in} \times 12 \text{ in} = 54 \text{ in}^2$, and the area of the actual room is $600 \text{ ft}^2 = 600 \text{ ft}^2 \times 144 \text{ in}^2/1 \text{ ft}^2 = 86,400 \text{ in}^2$. The ratios of the two areas is $86,400 \text{ in}^2/54 \text{ in}^2 = 1600$. The length of the actual room is $15 \text{ feet} = 15 \text{ ft} \times 12 \text{ in/ft} = 180 \text{ in}$, and the ratio of the lengths is $180 \text{ in}/4.5 \text{ in} = 40$. What we notice is that $1600 = 40^2$. In general, *the ratio of the areas is the square of the ratios of the lengths*. This fact is useful to remember.

Travel

From the formula for motion

$$\text{Distance} = \text{rate} \times \text{time}$$

we have the following equivalent formula.

$$\text{Time} = \frac{\text{distance}}{\text{rate}}$$

In a problem where the time taken is the same for similar modes of transportation, we can set up a proportion between the distances and rates.

Example 9

Test drives of a new automobile indicated that while traveling at two speeds, one of which was 20 mph greater than the other, the higher speed covered a test distance of 1000 ft while the lower speed covered a distance of 800 ft in the same time. What were the tested speeds, and what was the test time to the nearest second?

Solution 9

Since the test times were the same, we can use the ratio of distance/rate for each situation and create a proportion to model the problem:

$$\frac{\text{Distance of slower test}}{\text{Slower rate}} = \frac{\text{distance of faster test}}{\text{faster rate}}$$

Before beginning, however, we need to convert the distances into miles since the rates are given in miles per hour.

$$\frac{1000 \text{ ft}}{1} \times \frac{1 \text{ mi}}{5280 \text{ ft}} = 0.189 = 0.19 \text{ mi}$$

$$\frac{800 \text{ ft}}{1} \times \frac{1 \text{ mi}}{5280 \text{ ft}} = 0.152 = 0.15 \text{ mi}$$

Let r be the slower rate and $r + 20$ be the faster rate. The model gives us the equation

$$\frac{0.15}{r} = \frac{0.19}{r + 20}$$

Cross-multiplying, we have $0.15(r + 20) = 0.19r$.

Multiplying by 100, we have $15(r + 20) = 19r$, and we can solve this equation in the usual way.

$$15r + 300 = 19r$$

$$300 = 4r$$

$$r = 75 \text{ mph}$$

The slower tested speed was 75 mph, while the faster tested speed was 95 mph. Calculating the time tested actually gives us the check on the problem since both distances and speeds should produce the same time.

$$\textit{Time of slower test} = \frac{\text{distance}}{\text{rate}} = \frac{0.15}{75} = 0.002 \text{ hour}$$

$$\textit{Time of faster test} = \frac{\text{distance}}{\text{rate}} = \frac{0.19}{95} = 0.002 \text{ hour}$$

To convert to seconds, we use

$$\frac{0.002 \text{ hour}}{1} \times \frac{60 \text{ minutes}}{1 \text{ hour}} \times \frac{60 \text{ seconds}}{1 \text{ minute}} = 7.2 = 7 \text{ seconds}$$

Direct Variation

Another way in which problems make use of proportions is to refer to a *direct variation* among changing quantities; that is, when one aspect of a situation changes, another aspect changes in a proportional way or at the same rate.

Example 10

A construction firm knows that the number of new houses that it can build in a month varies directly with the cost of labor due to overtime charges. If 12 houses can be built when it pays an average of $14.50 per hour, how much should the company expect to pay its workers, if it needs to build 20 houses?

Solution 10

Since we are told that the quantities vary directly, we can immediately set up the model proportion.

$$\frac{\text{Number of houses}}{\text{Cost of labor}} = \frac{12}{4.50}$$

Let c be the cost of labor we are seeking. The model gives us

$$\frac{20}{c} = \frac{12}{14.50}$$

Cross-multiplying, we have $12c = 290$ and $c = 290/12 =$ $24.17 per hour.

Check: The check would be that the ratios would be the same, that is, $12 \div 14.50 = 0.83$ and $20 \div 24.17 = 0.83.$✓ (Note that the calculator shows differences to the fourth decimal place. This is due to the rounding we did in the problem.) The ratio 0.83 is referred to as the *constant of variation*.

Example 11

The force applied to stretch an elastic spring varies directly with the amount that it is stretched. If a force of 15 dynes (the metric measure of force) stretches a spring 2.3 cm, how much force, to the nearest dyne, would be required to stretch the spring 6 cm?

Solution 11

    Since we are told that the quantities vary directly, we can immediately set up the model proportion:

$$\frac{\text{Amount of force}}{\text{Stretch}} = \frac{15}{2.3}$$

Let f be the required force, and we have

$$\frac{f}{6} = \frac{15}{2.3}$$

and $f = 90/2.3 = 39.13 = 39$ dynes.

Check: The constant of variation must be the same: $15 \div 2.3 = 6.5$ and $39 \div 6 = 6.5$. ✓

Percentage

Problems that involve percentages are really problems involving proportions. When asked for a percentage of a number, you can think of it as asking "How much would we have if we scaled our base amount to be 100?"

    For example, when asked "What is 35% of 80?" we can translate this to mean "If we have 35 objects out of a set of 100, how much would we have in a *similar* group of 80?" The problem is then easily modeled by the proportion $35/100 = x/80$. This gives us $100x = 35 \times 80 = 2800$ and $x = 28$. Percentages appear in many of the problem situations we have seen before.

Investment

Example 12

When Gary invested $1200 in the Lion Fund, he had $1240 at the end of 3 months. Assuming the same rate of return, what would be the annual rate of return for this fund?

Solution 12

    In order to determine the annual rate, we have to make use of the given assumption and conclude that over the year

78

he would earn 4 times the interest received, since 3 months is one-fourth of a year. In other words, $4 \times \$40 = \$160$. Using the percentage model, we have

$$\frac{160}{1200} = \frac{x}{100}$$

Cross-multiplying gives us $1200x = 16,000$ and $x = 16,000/1200 = 13^{1}/_{3}\%$.

Check: We can calculate the interest for one year at $13.33\ldots$ percent and, then multiply by 0.25 to determine what the quarterly interest is; that is, $\$1200 \times 0.133333 \times .25 = \$39.9999$, which is approximately \$40. (The difference arose because we did not have to use the entire repeating decimal.)✓

Area

Example 13

The floor of a social hall that is 45 ft by 82 ft is carpeted except for a circular dance floor in the middle that has a diameter of 20 ft. To the nearest tenth of a percent, what percent of the floor is carpeted?

Solution 13

The solution requires finding the area of the entire floor, the area of the circle, and their difference, which is the amount of carpeted space. The percentage model is then used to give us

$$\frac{\text{Area of carpeted space}}{\text{Area of entire room}} = \frac{x}{100}$$

The area of the entire room is $45 \times 82 = 3690$ ft^2. The radius of the dance floor is 10 ft and using 3.14 for π, the area of the circular dance floor is $3.14 \times 10^2 = 314$ ft^2. Therefore, the carpeted area is $3690 - 314 = 3376$ ft^2.

The model gives us the proportion

$$\frac{3376}{3690} = \frac{x}{100}$$

Solving for x, we have $x = 3376 \times 100/3690 = 91.49 = 91.5\%$.

Check: The check is simple: $3690 \times 0.915 = 3376$ (rounded).✓

More complicated problems involving percentages can be found in *mixture problems* involving liquids that contain dissolved ingredients such as chemicals.

Example 14

A certain chemical solution used in manufacturing batteries contains water and acid. Initially, the solution is made with 50 kg of dry acid and 900 L of water. The solution is mixed and left to sit so that the water evaporates until it becomes a 30% acid solution. How much water, to the nearest liter, has to evaporate for this to be accomplished?

Solution 14

First, we need to make sure that our units of measurement are the same. Fortunately, in the metric system, 1 L of water weighs exactly 1 kg.

Regarding 30% as 30/100, the model for the problem is the proportion

$$\frac{\text{Amount of acid}}{\text{Remaining water} + \text{amount of acid}} = \frac{30}{100}$$

Let x be the amount of water that has evaporated. Therefore, $900 - x$ is the amount of remaining water and the total weight, in kilograms, is $900 - x + 50 = 950 - x$. The model gives us the equation:

$$\frac{50}{950 - x} = \frac{30}{100}$$

Cross-multiplying, we have $30(950 - x) = 5000$.
Simplifying and solving, we have

$$28{,}500 - 30x = 5000$$
$$23.500 = 30x$$
$$x = 23{,}500/30 = 783\tfrac{1}{3}$$

Therefore, 783 L of water has to evaporate.

80

Check: The remaining water is $900 - 783 = 117$, and the total weight is 50 kg of acid $+ 117$ kg of water $= 167$ kg, that is, $50/167 = .299 = 29.9\%$, which is okay since we used a rounded amount.✓

Summary

1. When given a ratio of two or more objects, assign a variable to be the multiple of each number in the ratio and use the specific multiples given by the ratio as the amounts. Model the problem and substitute these variable amounts to derive an equation to solve. Solve for the multiple and compute the actual amounts.
2. Identify when a proportion is implied in the problem. This will be the case when phrases such as "in the same ratio as," "varies directly as," or "is proportional to" appear in the problem.
3. Shapes are similar only when corresponding sides are in proportion.
4. Computing a percentage or using a given percentage will usually be solved by creating a proportion in which the ratio of the actual amounts is equal to $x/100$, where x is the percentage.

Additional Problems

1. Separate 133 into two parts so that the ratio of the larger to the smaller is 4:3.
2. On a box for a model-airplane kit, labels indicated that the scale used is 1:72. If the wingspan of the model is 9 in, how long is the wingspan for the actual plane to the nearest foot?
3. CD Emporium finds that CDs of rock artists outsold CDs of classical music by 5:2 during November. If the profit from a rock CD is $2.50 and the profit from a classical CD is $1.80, how much profit was made if 2345 rock and classical CDs were sold during November?
4. A survey to determine attitudes of students of mathematics was given to 420 sixth-grade students 5 years ago and to the same group now, when they are in eleventh grade. In the sixth grade, the ratio of students who found math easy to those who didn't was 3:1; now, in the eleventh grade, it is 3:2. How many fewer students find math easy now than before?
5. Glenn High School has 1620 students in which the girl:boy ratio is 5:4. Of this group, 8% of the girls and 5% of the boys went on a field trip. What percent, to the nearest tenth, of the school's population went on this trip?

6. The ratio of the longer side of a rectangle to the shorter side is 7:2. The ratio of another rectangle's longer side to its shorter side is 4:3. Given that the longer side of the second rectangle is 11 in more than the longer side of the first and the shorter side is 18 in more than the shorter side of the first rectangle, find the ratio of the perimeters of the first rectangle to the second.

7. On a scale model of an outdoor patio at the home design center, there is an equilateral triangle in its center that is to be filled with a specially colored concrete mix costing $15.25 per bag, and each bag can be used to cover 8 ft^2. The model is scaled at a ratio of 1:8 compared to the actual size of the patio, and a side of the triangular region is 1.5 feet. How much will it cost to fill the region on the actual patio?

8. The sum of the digits of a two-digit number is 12. If the digits are reversed, a larger number is formed and the ratio of the larger number to the smaller number is 7:4. Find both numbers.

9. The ratio of the ages of a girl, her mother, and the girl's grandmother is 2:5:8. Six years ago the ratio of the girl's age to her mother's age was 4:13. How old are the three women now?

10. Carl's crew can complete 7 jobs in the same time that Ben's crew can complete 5 jobs. During the same week, Carl's crew completed 12 more jobs than Ben's crew. How many jobs were completed by both crews?

11. Fifty liters of an acid solution contains 18 L of pure acid. How many liters of acid must be added to make a 4% acid solution?

12. The amount of calories in a serving of ice cream varies directly with the amount of sugar the ice cream contains. A company makes a regular chocolate ice cream and a "Lite" version. Regular chocolate ice cream has 240 calories (cal) per serving and contains 64 g of sugar. How many grams of sugar does the Lite chocolate have, if it has only 180 cal per serving?

13. Mrs. Simmons tries to be a shrewd investor and invested her $28,000 in three different stocks in the ratio of 3:4:7. Over the course of a year, the respective returns were 8, 12, and 10%. What was her overall percentage return for the year?

14. A plane can make the same 1000-mi trip as an express train in 6 hours' less time. The ratio of their speeds is 5:2. At what speed does each vehicle travel?

Solutions to Additional Problems

1. Let x be the larger part of 133. Therefore, $133 - x$ is the smaller part. The problem gives us the exact proportion $x : 133 - x = 4 : 3$. Using the product of the means is equal to the product of the extremes, we have

$$4(133 - x) = 3x$$
$$532 - 4x = 3x$$
$$532 = 7x$$
$$x = 532/7 = 76$$

   The larger part is 76, and the smaller part is $133 - 76 = 57$.

   Check: $76 + 57 = 133$ and $76/57 = 1.333\ldots = 1\frac{1}{3} = \frac{4}{3}.\checkmark$

2. This is a simple problem that will use the following proportion as its model.

$$\frac{\text{Wingspan of model}}{\text{Actual wingspan}} = \frac{1}{72}$$

   We can find the actual length in inches and then convert to feet. Let x be the number of inches in the actual wingspan. Therefore, we have $9\text{ in}/x = 1/72$. Cross-multiplying, we have $x = 9 \times 72 = 648$ in.

$$\frac{648\text{ in}}{1} \times \frac{1\text{ ft}}{12\text{ in}} = 54\text{ ft}$$

   Check: We need to make sure that the ratio of the actual span to the model span is 1:72 or 0.0139. Using the calculated inches, we have $9 \div 648 = .0139.\checkmark$

3. We need to first find the amount of each type of CD sold. The total profit is the sum of the profits from each type. The model for the problem consists of two steps.

   Step 1 Number of rock CDs + the number of classical CDs = 2,345.

   Step 2 Total profit = the number of rock CDs × \$2.50 + the number of classical CDs × \$1.80.

Let x = multiple of 2 and 5

$2x$ = number of rock CDs sold in November

$5x$ = number of classical CDs sold in November

From the model for step 1

$$2x + 5x = 2345$$
$$7x = 2345$$
$$x = 2345 \div 7 = 335$$

Therefore, $2 \times 335 = 670$ classical CDs were sold and $5 \times 335 = 1675$ rock CDs were sold. From the model for step 2

$$\text{Total profit} = 1675 \times \$2.50 + 670 \times \$1.80$$
$$= \$4187.50 + \$1206.00 = \$5393.50$$

Check: The check is concerned mostly with the amounts of each type of CD sold. We need to make sure that the ratio is indeed 5:2 or 2.5, specifically, $1675 \div 670 = 2.5$.✓ It is also necessary to recheck the arithmetic for step 2.

4. We have to compute the actual numbers of students who find math easy and those who do not at each grade level on the basis of the given ratios. For the sixth grade, let x be the multiple, $3x$ be the number of students who found math easy, and $1x$ be the number of students who did not.

$$3x + x = 420$$
$$4x = 420$$
$$x = 105$$

Therefore, at the sixth grade, there were $3 \times 105 = 315$ students who found math easy (and 105 who did not). At the eleventh grade (we should use a different variable so to avoid confusion between our two answers),

Let y = multiple

$3y$ = number of students who find math easy

$2y$ = number of students who do not

$$3y + 2y = 420$$
$$5y = 420$$
$$y = 84$$

Therefore, at the eleventh grade there are $3 \times 84 = 252$ students who find math easy (and $2 \times 84 = 168$ who do not). The difference $315 - 252 = 63$ is our answer.

Check: The check involves making sure that the numbers we determined are in the ratio given in the problem. The sixth-grade ratio should be 3 : 1 or 3, that is, $315 \div 105 = 3$. ✓ The eleventh-grade ratio should be 3:2 or 1.5, that is, $252 \div 168 = 1.5$ ✓

5. The problem must be solved by finding the number of girls and boys who went on the trip, using the percentage model

$$\frac{\text{Number of girls on trip} + \text{number of boys on trip}}{\text{Total number of students}} = \frac{x}{100}$$

Realizing this, we see that simply adding 8% + 5% = 13% is most likely *not* the correct answer. (This is the most common error in a problem like this.) It would be helpful to have an estimate before beginning. The ratio 5:4 indicates that slightly more than half of the students are girls. If the groups were equal, there would be 810 girls and 810 boys. Using 10% as an estimate for 8%, about 80 girls would be on the trip, and 5% of the boys would mean 40 boys on the trip. An estimate of the total number of students on the trip would be 120 and $120 \div 1620 = 0.074$, which is approximately 7 percent.

The algebraic scheme for finding the ratios will give us the numbers of girls and boys in the school.

$$\text{Let } x = \text{multiple}$$
$$5x = \text{number of girls}$$
$$4x = \text{number of boys}$$
$$5x + 4x = 1620$$
$$9x = 1620$$
$$x = 180$$

There are $5 \times 180 = 900$ girls and $4 \times 180 = 720$ boys. Thus 8% of the girls is $900 \times 0.08 = 72$ and 5% of the boys is $720 \times 0.05 = 36$. Therefore, $72 + 36 = 108$ students went on the trip. The model gives us the proportion: $108/1620 = x/100$ and $x = 10,800 \div 1620 = 6.66\ldots = 6.7\%$ rounded to the nearest tenth.

Check: The best check for a problem with these many steps is to recheck the arithmetic at each step and to feel sure that the answer is reasonable. The answer is reasonable on the basis of our initial estimate.

6. Since the only information about the first rectangle is the ratio of the sides, we will need to use the information about the second rectangle to help find the sides. Let x be the multiple needed for the first rectangle, $7x$ be the longer side of the first rectangle, and $2x$ be the shorter side. Then $7x + 11$ is the longer side of the second rectangle and $5x + 18$ is the shorter side of the second rectangle. We can set up the following proportion for the second rectangle.

$$\frac{7x + 11}{2x + 18} = \frac{4}{3}$$

Cross-multiplying, we have the equation $3(7x + 11) = 4(2x + 18)$. Simplifying and solving, we obtain

$$21x + 33 = 8x + 72$$

$$13x = 39$$

$$x = 3$$

Therefore, the dimensions of the first rectangle are $7 \times 3 = 21$ in and $2 \times 3 = 6$ in. Its perimeter is $2 \times (21 + 6) = 54$ in. The sides of the second rectangle are $21 + 11 = 32$ in and $6 + 18 = 24$ in. Its perimeter is $2 \times (32 + 24) = 112$ in. The ratio we want is $54{:}112$ or $54/112$, which reduces to $27/56$ or $27{:}56$.

Alternative Solution   After having assigned the variable and obtained representations for the sides, we could have set up our answer in variable form immediately. Specifically, the perimeter of the first rectangle is $2(7x + 2x) = 18x$ and the perimeter of the second rectangle is $2(7x + 11 + 2x + 18) = 18x + 58$. The ratio of the perimeters would be $18x/(18x + 58)$. Once we found that $x = 3$, from the proportion, the ratio is quickly seen as 54/112.

7. The ratio plays only a small part in this problem. A scale of 1:8 simply means that the actual patio will be 8 times larger than the model. Therefore, the actual side of the triangle will be $8 \times 1.5 = 12$ ft. Using the formula for the area of an equilateral triangle from the reference table in Chap. 1, we find that the area is $\sqrt{3}/4 \times 12^2$ or approximately $1.73 \div 4 \times 144 = 62.28$ ft^2. Since each bag covers 8 ft^2, we will need $62.28 \div 8 = 7.785$ bags, which means that we have to buy 8 full bags. The cost is $15.25/bag $\times$ 8 bags $=$ $122.00.

8. A simple way to solve the problem is to try all possibilities. The only two-digit numbers whose sum of digits is 12 and which are smaller than the number whose digits are reversed are 39, 48, and 57. The ratio 7:4 is to equivalent to 1.75. The ratios to consider are 93/39, 84/48, and 75/57. The only one of these equal to 1.75 is 84/48. Therefore, the numbers are 48 and 84.

An algebraic solution to problems involving the digits of a number are best solved by using two variables, one for the tens digit and one for the units digit. We know that the original number is the smaller of the numbers. Let u be the units digit and t be the tens digit of the original number. Therefore, $10t + u$ is the original number and $10u + t$ is the number when the digits are reversed. The problem gives us two pieces of information to use: (a) $t + u = 12$ and (b) $(10u + t)/(10t + u) = 7/4$. Using ($b$) and cross multiplying, we have $4(10u + t) = 7(10t + u)$. Simplifying gives us $40u + 4t = 70t + 7u$. From (a) we see that $t = 12 - u$ and when substituted into the equation, we have an equation in one variable to solve

$$40u + 4(12 - u) = 70(12 - u) + 7u$$
$$40u + 48 - 4u = 840 - 70u + 7u$$
$$36u + 48 = 840 - 63u$$
$$99u = 792$$
$$u = 792/99 = 8 \text{ (and } t \text{ must be } 12 - 8 = 4)$$

Therefore, the original number is 48 and the larger number is 84.

Check: We need to ensure that the ratio of the numbers is $7/4 = 1.75$, or $84/48 = 1.75.\checkmark$

9. An algebraic solution is the fastest way to the answer. Let x be the multiple for the current ages, $2x$ be the girl's age, $5x$ be the mother's age, and $8x$ be the grandmother's age. Six years ago, the girl's age was $2x - 6$ and the mother's age was $5x - 6$. With these, we have the proportion

$$\frac{2x - 6}{5x - 6} = \frac{4}{13}$$

Cross multiplying gives us the equation $13(2x - 6) = 4(5x - 6)$. Simplifying and solving, we have

$$26x - 78 = 20x - 24$$
$$6x = 54$$
$$x = 9$$

Currently, the girl is $2 \times 9 = 18$ years old, the mother is $5 \times 9 = 45$ years old, and the grandmother is $8 \times 9 = 72$ years old.

Check: Six years ago, the girl and her mother were 12 and 39 years old, respectively. The quotient ratio 12/39 is reducible to 4/13.✓

10. The first sentence of the problem gives us the ratio

$$\frac{\text{Jobs completed by Carl's crew}}{\text{Jobs completed by Ben's crew}} = \frac{7}{5}$$

Let x be the multiple, $7x$ be the number of jobs completed by Carl's crew, and $5x$ be the number of jobs completed by Ben's crew during the week. The second sentence gives us the simple model

Number of jobs completed by Carl's crew =
number of jobs completed by Ben's crew $+ 12$

Inserting the variables, we have the equation $7x = 5x + 12$ and, clearly, $x = 6$. Therefore, Carl's crew completed $7 \times 6 = 42$ jobs and Ben's crew completed $5 \times 6 = 30$ jobs. The total number of jobs was 72.

Check: We need to ensure that 42:30 is equivalent to 7:5. We find that 42/30 does reduce to 7/5.✓

11. To better understand the problem, it is helpful to realize that the original solution is 36 percent acid; that is,

$$\frac{\text{Amount of acid}}{\text{Amount of acid} + \text{water}} = \frac{18}{50} = \frac{36}{100} = 36\%$$

Since we are adding more acid, we let x be the amount of acid added. This enables us to create the proportion $(18 + x)/(50 + x) = 40/100$. Cross-multiplying gives us the equation $100(18 + x) = 40(50 + x)$. Simplifying and solving for x, we obtain

$$1800 + 100x = 2000 + 40x$$
$$60x = 200$$
$$x = 3\tfrac{1}{3} \text{ L}$$

Check:

$$\frac{18 + 3\tfrac{1}{3}}{50 + 3\tfrac{1}{3}} = \frac{21\tfrac{1}{3}}{53\tfrac{1}{3}} = \frac{64}{3} \div \frac{160}{3} = \frac{64}{3} \times \frac{3}{160}$$
$$= \frac{64}{160} = 0.4 = 40\%\checkmark$$

12. The phrase "varies directly" indicates that the ratios of calories to amount of sugar are proportional. Therefore, letting g be the number of grams of sugar in the Lite version, we have the proportion $240/64 = 180/g$. Cross multiplying gives us $240g = 180 \times 64 = 11{,}520$ and $g = 11{,}520 \div 240 = 48$ g of sugar.

Check: $240/64 = 3.75$ and $180/48 = 3.75.\checkmark$

13. Note the word *respective*. This is usually meant as a direction to make a correspondence between the first items in each list, the second items in each list, and so on. In this problem, it means that the smallest investment earned 8%, the middle amount earned 12%, and the largest investment earned 10%. Each percentage return must be calculated for the money in that investment. Therefore, we have to solve this problem in several steps.

(*a*) Find the amount of each investment. Let x be the multiple. The three amounts invested are $3x$, $4x$, and $7x$. Therefore, $3x + 4x + 7x = 14x = 28{,}000$ and $x = 2000$. The three amounts are actually $6000, $8000, and $14{,}000.

(b) Find the new amounts on the basis of the percentage returns. We can use the formula $A = p + prt$, where $t = 1$. For the $6000 investment, she earned $6000 \times .08 = \$480$. For the $8000 investment, she earned $8000 \times .12 = \$960$. For the $14,000 investment, she earned $14,000 \times .10 = \$1400$. The total return was $\$480 + \$960 + \$1400 = \$2840$.

(c) Find the overall percentage using $2840/28,000 = x/100$, that is, $x = 2840 \times 100 \div 28,000 = 10.14\%$.

Check: The most important aspect of the problem to check is the amounts of each investment represented in the ratio of 3:4:7. Checking for a ratio involving three or more terms can be done pairwise: $\$8000/\$6000 = 4/3$, $\$14,000/\$8000 = 7/4$. Both are true. The rest of the check would be to recheck the arithmetic in finding the earnings and overall percentage.✓

14. We can use two variables to set up this problem quickly. Let x be the multiple for the speeds. Since the plane is obviously the faster vehicle, $5x$ is the speed of the plane and $2x$ is the speed of the train. Let t be the time of the train and, therefore, $t - 6$ is the time of the plane. Using Distance = rate × time, we have $2xt = 1000$ or $xt = 500$, and realizing that the distances traveled by each are equal, we have the equation $2xt = 5x(t - 6)$. Simplifying gives us $2xt = 5xt - 30x$ or $30x = 3xt$. Using $xt = 500$ from above, we have $30x = 1500$ and $x = 50$.

Therefore, the train travels at $2 \times 50 = 100$ mph and the plane travels at $5 \times 50 = 250$ mph.

Check: We need to check the ratios of the speeds and the time taken by each vehicle. The quotient ratio $250/100$ can be reduced to $5/2$.✓ Using Time = distance ÷ rate, the time taken by the train is 1000 mi $÷ 100$ mph $= 10$ hours, and the time taken by the plane is 1000 mi $÷ 250$ mph $= 4$ hours, which is 6 hours less.✓

Chapter 4

Word Problems Involving Geometry and Trigonometry

Geometry, the study of the relationships among shapes, is a major component of mathematics as it enables us to understand the physical world around us. *Trigonometry* is the study of the relationships within triangles and, therefore, the two areas are closely related. Word problems that involve geometry and trigonometry mostly require an understanding of a few basic relationships between lines, angles, and polygons. The key step is to identify the relationship that applies. Many of the basic relationships are listed at the end of the chapter in the Table of Fundamental Geometric and Trigonometric Relationships; others are mentioned in the solutions to the problems as they arise. Algebraic equations will arise, and you should refer to Chap. 2 and to the Appendix for review and practice if you have difficulty following the steps in their solutions.

Diagrams are essential in problems involving shapes. In most cases, the shapes are simple to draw. At other times the problems describe the geometric figure in detail in order to test your ability to visualize the situation. This requires careful reading and following the description.

Problems Involving Angles

Example I

The ratios of the measures of the angles of a triangle are 2:7:9. Find the measure of the largest angle.

Solution 1

The model for the problem comes from the fact that the sum of the measures of the angles of a triangle is 180°.

It is nearly obvious that the angle measures 20°, 70°, and 90° solve the problem. To be sure, we make use of the method seen in Chap. 3 to solve ratio problems. A diagram is also helpful in understanding the problem.

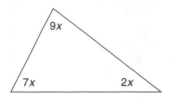

Let x be the multiple. The angles are, therefore, $2x$, $7x$, and $9x$. The model gives us the equation

$$2x + 7x + 9x = 180$$

$$18x = 180$$

$$x = 10$$

The angles are indeed $2 \times 10 = 20°$, $7 \times 10 = 70°$, and $9 \times 10 = 90°$.

Example 2

Two cables of equal length are supporting a vertical antenna on a roof. Each is attached to the top of the antenna and to a spot on the rooftop. If the cables form a 40° angle with each other, what are the measures of the angles formed by the cables and the rooftop?

Solution 2

A diagram is necessary to visualize the situation and identify the shape involved.

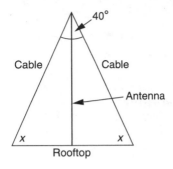

Since the cables are of equal length, the triangle has equal sides and is, therefore, *isosceles*. In an isosceles triangle the angles opposite the congruent sides are congruent. Using this along with the fact that the sum of the measures of the angles of a triangle is 180°, we can derive an equation to solve.

Let x be the measure of each of these congruent angles.

$$x + x + 40 = 180$$
$$2x + 40 = 180$$
$$2x = 140 \quad \text{and} \quad x = 70°$$

Many word problems involve a pair of parallel lines. The relationships that arise with parallel lines concern the angles formed when a third line is introduced that intersects both parallel lines. This line is called a *transversal*. The next example demonstrates such a situation.

Example 3

The largest of the angles formed by a transversal that crosses two parallel lines is 45° less than twice the smallest of the angles. Find the measures of these angles.

Solution 3

All the facts about the angles that are formed when a transversal crosses two parallel lines indicate that only two different angle measures appear.

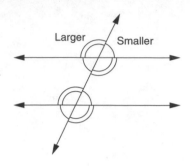

Larger   Smaller

Therefore, the "largest" angle is the larger of these two measures and the "smallest" is the smaller of these measures. These are clearly adjacent angles that lie on the same straight line and are, therefore, supplementary. The model uses this fact and the following relationship in the problem:

Measure of larger angle + measure of smaller angle = 180°

Measure of larger angle = 2 × measure of smaller angle = 45°

Since the problem relates the larger angle to the smaller angle, we will represent the smaller angle by the variable.

$$\text{Let } x = \text{smaller angle}$$
$$2x - 45 = \text{larger angle}$$
$$x + 2x - 45 = 180$$
$$3x = 225$$
$$x = 75$$

The smaller angle is 75 and the larger angle is $2 \times 75 - 45 = 105°$.

Check: The angles must be supplementary: $75° + 105° = 180°.\checkmark$

Problems Involving Line Segments of Polygons

Many relationships exist among the line segments found in triangles, quadrilaterals, and other polygons depending on the

94

type of figure in question. These line segments could be sides of the figure, diagonals, altitudes, medians, angle bisectors, and others. The table at the end of the chapter lists many of these relationships. The most common word problems involve similar figures that are found in the diagram that models the problem.

Example 4

At some time in the afternoon a 6-ft person casts a shadow that is 10 ft long while a nearby flagpole casts a shadow that is 22 ft long. To the nearest foot, how tall is the flagpole?

Solution 4

In order to identify the shapes we are working with, we need to draw a diagram. Some basic assumptions must be made: (1) the person and the flagpole are standing perpendicular to level ground and (2) the sun's rays form the same angles with the head of the person and the top of the flagpole. These assumptions are standard for this kind of problem and lead to the following diagram.

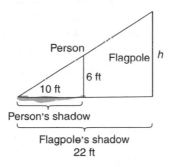

Note that two similar right triangles are formed on the basis of the assumptions. Since corresponding sides of similar triangles are in proportion, we have

$$\frac{\text{Height of person}}{\text{Length of person's shadow}} = \frac{\text{height of flagpole}}{\text{length of flagpole's shadow}}$$

Let h be the height of the flagpole, and we have $6/10 = h/22$. Cross-multiplying gives us $10h = 132$ and $h = 13.2$. The height of the flagpole to the nearest foot is 13 ft.

95

The next example demonstrates the need to carefully read the description of the shape involved in the problem. Part of the goal of such a problem is to test your ability to visualize the situation. This requires careful reading and following the description.

Example 5

One of the shorter sides of a rectangle is extended 10 units beyond a vertex, and a line segment is drawn from its new endpoint through the rectangle to the vertex opposite the vertex extended through. This segment cuts the side it is drawn through into two segments of 15 and 24 units. What is the area of the rectangle?

Solution 5

Finding the area of the rectangle requires finding the length and width. The problem tells us that one side is $24 + 15 = 39$ units. We need to study a diagram of the problem in order to find the means to find the width.

The directions given in the problem must be followed with care in order to arrive at the correct diagram. The vertices of the starting rectangle ought to be labeled, and as the directions are followed new labels should mark points of intersection.

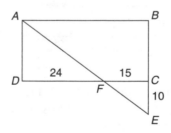

Suppose that we extend side BC through C to a new endpoint E. The problem tells us that $CE = 10$ units. The vertex opposite C is A, and we draw segment AE, which cuts side DC at F. The problem tells us that DF is 24 and FC is 15. The diagram makes it clear that there are two

96

similar right triangles, $\triangle ADF$ and $\triangle ECF$. Since corresponding sides of similar triangles are in a proportion, we have $AD/EC = DF/CF$ or $AD/10 = 24/15$. Thus $AD = 240 \div 15 = 16$ units. The area of the rectangle is $39 \times 16 = 624$ square units.

Problems Involving Right Triangles

The most frequent real-world applications of geometry include situations that involve right triangles. This is due to finding right triangles in abundance in the physical structures of the world around us and the importance of finding perpendicular distances. The two most important tools in right-triangle problems are *the Pythagorean Theorem* and the trigonometric ratios: *sine, cosine,* and *tangent.*

Using the Pythagorean Theorem

The Pythagorean Theorem is used when we know two sides of the right triangle and need to find the third. It relates the two legs and the hypotenuse of a right triangle in the following way: *The square of the length of the hypotenuse is equal to the sum of the squares of the lengths of the legs.* Using variables a and b to represent the lengths of the legs and c to represent the length of the hypotenuse, we have the familiar statement $a^2 + b^2 = c^2$.

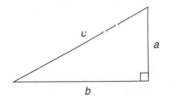

Remember that the hypotenuse is always the side opposite the right angle and is the longest side of the right triangle.

In solving problems with the Pythagorean Theorem, you will have to compute square roots. These are seldom going to be whole numbers. Therefore, be sure to give the answers to the degree of accuracy asked for in the problem.

Example 6

A 30-ft ladder leans against the side of a building and reaches the flat roof. If the ladder stands at a point 10 ft from the house, how tall is the house to the nearest foot?

Solution 6

A simple diagram of the situation clearly reveals that we are working with a right triangle in which we know two sides and have to determine the third.

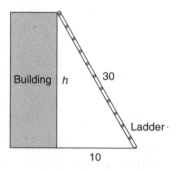

More specifically, we know one leg and the hypotenuse. If we let h be the height of the building, we have the equation $h^2 + 10^2 = 30^2$.

$$h^2 + 100 = 900 = 800$$
$$h = \sqrt{800} = 28.28$$

The height of the building to the nearest foot is 28 ft.

Several sets of numbers, referred to as *Pythagorean Triples*, are useful to remember. Each triple is a set of three whole numbers that satisfy the Pythagorean Theorem. The four most common triples are

$$3, 4, 5 \qquad 3^2 + 4^2 = 9 + 16 = 25 = 5^2$$
$$5, 12, 13 \qquad 5^2 + 12^2 = 25 + 144 = 169 = 13^2$$
$$8, 15, 17 \qquad 8^2 + 15^2 = 64 + 225 = 289 = 17^2$$
$$7, 24, 25 \qquad 7^2 + 24^2 = 49 + 576 = 625 = 25^2$$

If you notice two of the three numbers in a problem, you can save computation time. Be sure that the smaller numbers represent the legs and the largest one represents the hypotenuse!

Actually, any triplet that is in the same ratio as any of the above will also satisfy the Pythagorean Theorem. For example

$$6, 8, 10 \qquad 6^2 + 8^2 = 36 + 64 = 100 = 10^2$$

Look for these scaled triplets as well.

Example 7

The sides of a triangle are 15, 15, and 18. What is the length of the altitude to the noncongruent side?

Solution 7

The altitude creates two congruent right triangles and, hence, it bisects the side. The following diagram indicates that we are looking for a leg of a right triangle whose other leg is 9 and whose hypotenuse is 15. A multiple of the 3:4:5 triple is 9:12:15. There is no other possibility for the leg but 12.

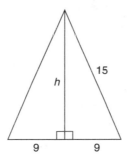

We can also solve the problem algebraically to be convinced of this. Let h be the length of the altitude. Therefore, $h^2 + 9^2 = 15^2$.

$h^2 + 81 = 225$ and $h^2 = 144$. Then $h = \sqrt{144} = 12$.

The Pythagorean Theorem shows up in other figures where right triangles are formed. Look for this when "dropping" altitudes or working with perpendicular lines.

Example 8

The parallel sides of an isosceles trapezoid are 8 and 18 units. If the area of the trapezoid is 156 square units, how long are the nonparallel sides?

Solution 8

The answer is found by dropping altitudes from the shorter of the parallel sides to the longer. The shape formed in the center is a rectangle, and the right triangles formed on its sides are congruent. Therefore, the longer base is divided into segments of 5, 8, and 5. Each nonparallel side of the trapezoid is a hypotenuse of the right triangle, in which one leg is 5 units and the other leg is the altitude of the trapezoid.

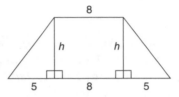

We can calculate the length of the altitude from the formula for the area of a trapezoid.

$$\text{Area} = \tfrac{1}{2} \times (\text{base } 1 + \text{ base } 2) \times h$$

We have $156 = \tfrac{1}{2} \times 26 \times h = 13h$. Therefore, $h = 156/13 = 12$. We now have a right triangle whose legs are 5 and 12. The hypotenuse must be 13 from the familiar 5,12,13 Pythagorean Triple.

Example 9

A floor pattern requires tiles that are in the shape of a rhombus. The manufacturer makes these tiles according to the lengths of the diagonals. How long is each side of the tile, if the diagonals are 10 and 16 in?

Solution 9

The solution to this problem lies in the fact that the diagonals of a rhombus are perpendicular. Since a rhombus is a parallelogram, the diagonals bisect each other. Therefore, the diagonals create four congruent right triangles in the interior, where each side of the rhombus is a hypotenuse of the right triangle.

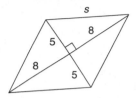

Letting s represent the side of the rhombus, the Pythagorean Theorem gives us the computation $s^2 = 5^2 + 8^2 = 25 + 64 = 89$ and $s = \sqrt{89}$. The problem doesn't ask for a specific degree of accuracy. It is best, therefore, to leave the answer in radical form. However, it is a good idea to get its value to see if the answer is reasonable. The square root $\sqrt{89}$ is approximately 9.43. This is reasonable since it is larger than either of the other two sides of the right triangle.

Using the Trigonometric Ratios

Trigonometric ratios are ratios of two sides of the right triangle and are used when we have problems that involve two sides and an acute angle of a right triangle. The ratios are referred to with respect to the angle in question and are called the *sine* of the angle, the *cosine* of the angle, and the *tangent* of the angle. The legs of the right triangle are identified as being either the side *opposite* the angle or the side *adjacent* to the angle. The other side is, of course, the *hypotenuse*. The ratios are specifically

$$\textit{Sine of an angle} = \frac{\text{length of opposite side}}{\text{length of hypotenuse}}$$

(*sine* is usually abbreviated as *sin*)

$$\textit{Cosine of an angle} = \frac{\text{length of adjacent side}}{\text{length of hypotenuse}}$$

(*cosine* is usually abbreviated as *cos*)

$$\textit{Tangent of an angle} = \frac{\text{length of opposite side}}{\text{length of adjacent side}}$$

(*tangent* is usually abbreviated as *tan*)

101

Letting the legs be represented by a and b and the hypotenuse by c as in the diagram below,

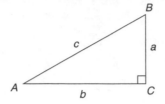

we have

$$\sin A = \frac{a}{c} \qquad \cos A = \frac{b}{c} \qquad \tan A = \frac{a}{b}$$

while

$$\sin B = \frac{b}{c} \qquad \cos B = \frac{a}{c} \qquad \tan B = \frac{b}{a}$$

Your calculator or a table of trigonometric values can tell you the approximate value of the sine, cosine, or tangent of any acute angle, and it also can tell you the angle if you know the ratio.

Finding Sides of a Right Triangle

Example 10

The warning on the side of a 25-ft ladder says that the angle it makes with the ground should be no smaller than 40° before it becomes unsafe. To the nearest foot, what is the height along a wall that this ladder makes at such an angle?

Solution 10

The diagram for the problem presents us with a right triangle. We should label the vertices of the triangle for reference.

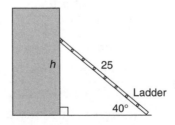

The side of the building is the side opposite the angle we are using, and the ladder is the hypotenuse. Therefore, we need to use the sine ratio. The verbal model is

$$\text{Sine of } 40° = \frac{\text{height along wall}}{\text{length of ladder}}$$

or using h for the height, $\sin 40° = h/25$. The calculator tells us that $\sin 40° = 0.6428$ (four decimal places is standard). Therefore, we have $h = 25 \times 0.6428 = 16.07$. To the nearest foot, our ladder will reach a height of 16 ft.

In word problems involving trigonometry, we often find the phrases "angle of elevation" and "angle of depression." In Example 10, the 40° angle that the ladder makes with the ground would be the angle of elevation of the ladder to the point where it reaches on the wall. In other words, the *angle of elevation* is the measure of the upward tilt off the horizontal required to view an overhead object. The *angle of depression* is the measure of the downward tilt off the horizontal to view an object below.

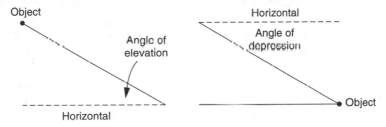

In both cases a right triangle is formed when drawing the height of the viewer or the object, and the trigonometric ratios can be used. It is important to note that the angle of depression lies outside the triangle and is the complement of its adjacent angle. Therefore, it is also equal to the other acute angle of the right triangle.

Example 11

From a point 150 ft from the center of the base of a hill, the angle of elevation to the top of the hill is 28°. A tram from that point carries people to the top. To the nearest foot, how high is the hill, and how long is the tram's cable?

Solution 11

The *angle of elevation* is the angle that the tram cable makes with the ground. The height of the hill is measured as the perpendicular axis from its top to the center of its base. The following diagram models the problem, and we label the vertices for reference. (Remember, you don't have to be an artist; use simple lines to make the right triangle easy to identify!)

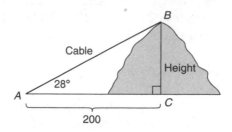

We know the length of the side adjacent to the angle $AC = 200$. The height is the side opposite the angle BC, and the length of the cable is the hypotenuse AB.

In order to find the height, we easily determine that the appropriate trigonometric ratio is tangent since it is the one of the three that involves the opposite and adjacent sides. Letting h be the height, we have $\tan 28° = BC/AC = h/200$. Therefore, $h = 200 \times \tan 28° = 200 \times 0.5317 = 106.34 = 106$ ft, to the nearest foot.

We can find the length of the cable in two different ways, one using trigonometry and the other using the Pythagorean Theorem.

- We can use the cosine ratio since it is the one that involves the adjacent side and the hypotenuse. Letting c be the length of the cable, we have $\cos 28° = AB/AC = 200/c$. When we cross multiply here, we have $c \times \cos 28° = 200$. Therefore $c = 200 \div \cos 28° = 200 \div 0.8829 = 227$ feet, to the nearest foot.

- We can use the Pythagorean Theorem to find the hypotenuse of the triangle, since we were given AC and we found BC. Note that we are using a value that we had to determine and we had rounded that answer. To be accurate, we should use the value of BC to more decimal places, 106.34, and work with several decimal places until we round off at the end of the calculations. Letting c be the cable length, we have

$$c^2 = 200^2 + 106.34^2 = 40{,}000 + 11{,}308.20 = 51{,}308.20$$

Therefore, $c = \sqrt{51{,}308.20} = 226.51$ ft $= 227$ ft, to the nearest foot. (If we would have used 106 ft for BC, our answer would have been 226 ft, which is close, but not correct.)

Example 12

An observer from the top of a lighthouse that is 40 ft tall and stands on a cliff 80 ft above sea level notices a boat off shore. Boats must be warned if they come too close to the cliff, because of the rocks below the waterline. If the observer notes that the angle of depression is 50°, how far from the cliff is the boat?

Solution 12

The diagram clearly shows the right triangle that must be used. It also indicates that the height includes the cliff and the tower, that is, $80 + 40 = 120$ ft.

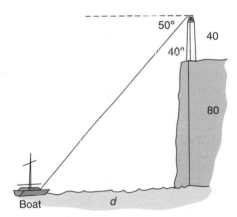

The diagram also indicates that the angle within the triangle we need to use measures 40°. The height of the cliff is the adjacent side of the angle, and the distance between the boat and the cliff is its opposite side. Therefore, we should use the tangent ratio. Letting d be the distance, we have $\tan 40° = d/120$ and $d = 120 \times \tan 40 = 120 \times 0.8391 = 100.7 = 101$ ft, to the nearest foot.

Finding Angles of a Right Triangle

The previous examples demonstrated how the trigonometric ratios are used to find the lengths of the sides of a right triangle. They are equally as useful when trying to find the acute angles of a right triangle. To accomplish this, you have to know the lengths of two sides. The important step is to identify the position of the given sides relative to the angle you seek to find.

Example 13

A young tourist visiting Washington, D.C. is awed by the size of the Washington Monument, which is 555 feet tall. The tourist, who is 5 ft tall, is standing 50 yards (yd) from the monument. At what angle, to the nearest degree, is she elevating her head when she looks at the top of the monument?

Solution 13

Two aspects of this problem must be accounted for before we can accurately identify the right triangle. One is the obvious change of units in the distance that the tourist is from the monument. She is standing 50 yd away, which is equal to 150 ft.

The second aspect is more subtle. Since we are concerned with the angle of elevation of her head, we must consider the horizontal axis to begin at the height of her head. We can safely say that is 5 ft. Therefore, the vertical height used is $555 - 5 = 550$ ft.

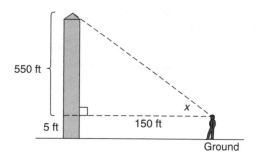

The monument is the side opposite the angle, and the distance is the side adjacent to the angle. We can make use of the tangent ratio to solve the problem.

106

Let x be the measure of the angle. We have $\tan x° = 550/150 = 3.6667$. Your calculator should have a key press that will find the *inverse tangent* (sometimes written as *tan*$^{-1}$ or *arc tan*). This will tell you what angle has the calculated number as its tangent. We could write $x =$ the inverse tangent of 3.6667 or $x = \tan^{-1} 3.6667$. Enter 3.6667, and press the key to get $x = 74.7°$ or, to the nearest degree, 75°.

Special Right Triangles

Just as there are some easy triplets of numbers to help with using the Pythagorean Theorem, certain special right triangles are well known and used frequently in problems involving trigonometry.

The most easily remembered trigonometric ratio is $\sin 30° = 0.5$ or $\frac{1}{2}$. This means that in a right triangle that has a 30° angle, the hypotenuse is twice the size of the side opposite the 30° angle. If these sides of the triangle were 1 and 2, the Pythagorean Theorem would be used to find that the other leg is $\sqrt{3}$.

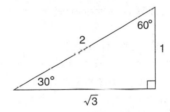

We refer to this as the special 30–60–90 *right triangle*.

From the diagram, we have six trigonometric ratios that are easy to remember and have exact radical values.

$$\sin 30° = \frac{1}{2} \qquad \cos 30° = \frac{\sqrt{3}}{2} \qquad \tan 30° = \frac{1}{\sqrt{3}}$$

$$\sin 60° = \frac{\sqrt{3}}{2} \qquad \cos 60° = \frac{1}{2} \qquad \tan 60° = \sqrt{3}$$

and we have the following ratio: leg opposite 30° : leg adjacent 30° : hypotenuse $= 1 : \sqrt{3} : 2$. This makes problems like the following easy to solve.

107

Example 14

The diagonal of a rectangle is 8 in long and makes a 30° with the longer side. What is the perimeter of the rectangle?

Solution 14

The diagram with the vertices labeled shows that we have a right triangle, $\triangle BAD$, with hypotenuse BD. Since the angle is 30°, it is a 30–60–90 right triangle and we know that $AD:AB:BD = 1 : \sqrt{3} : 2$.

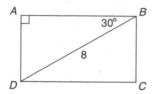

Therefore, since $BD = 8$ in we have $AD = 4$ in and $AB = 4\sqrt{3}$ in. The perimeter of the rectangle is $4 + 4 + 4\sqrt{3} + 4\sqrt{3} = 8 + 8\sqrt{3}$. This is an exact answer and does not require a decimal approximation unless one is asked for. (Note that $\sqrt{3}$ is approximately 1.73 if it is needed.)

The other special right triangle is the isosceles right triangle. Since the acute angles must be equal, they are both 45°. Since the legs are equal, their ratio is 1. Finally, since the legs are the sides opposite and adjacent to a 45° angle, we have that $\tan 45° = 1$! The Pythagorean Theorem gives us the hypotenuse exactly as $\sqrt{2}$.

The diagram below depicts the special 45–45–90 right triangle.

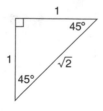

We have the three trigonometric ratios $\sin 45° = 1/\sqrt{2}$, $\cos 45° = 1/\sqrt{2}$, and $\tan 45° = 1$, and the sides are in the ratio leg:leg:hypotenuse $= 1 : 1 : \sqrt{2}$.

108

Example 15

What is the exact area of a square whose diagonal is 8 in?

Solution 15

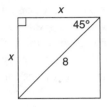

The diagonal creates two isosceles right triangles within the square.

Using the ratio leg:hypotenuse $= 1/\sqrt{2}$, if we let x be the side of the square, we have the proportion $x/8 = 1/\sqrt{2}$. Cross-multiplying gives us $x\sqrt{2} = 8$ and $x = 8/\sqrt{2}$ inches as the side of the square. The area of the square is, therefore, $(8/\sqrt{2})^2 = 64/2 = 32$ in^2.

Summary

1. Make sure to draw simple but accurate diagrams.
2. Look for right triangles, isosceles triangles, parallel lines, and other shapes that have known relationships.
3. Assign a variable to the unknown length or angle, and create a model equation that uses the identified relationship.
4. Solve the equation for the variable, and be sure to use it to determine all the lengths or angles asked for in the problem.
5. Check your arithmetic and algebra, and make sure that the answer is reasonable for the problem.

Additional Problems

1. In a triangle, two of the angles are in a ratio of 5:3. The third angle is 40° less than the smaller of the other two. Find the measure of the largest angle.
2. The acute angles of a right triangle are such that the square of the smaller is 25° less than 10 times the larger. Find both angles.
3. Two angles formed on the same leg of a trapezoid are such that one is 30° less than 5 times the other. Find the measures of the angles.

4. In an isosceles triangle, a segment drawn from the vertex angle to the base cuts the vertex angle into two angles whose ratio is 5:3. If the smaller of these angles is 20° less than the base angle of the isosceles triangle, what are the measures of the angles formed by the line segment?

5. An entrance to a mine is dug horizontally into a hill. At a point 45 m along the entrance, a ventilating shaft is constructed vertically upward and breaks through at a point 80 m along the hillside. How tall is the ventilating shaft, to the nearest meter?

6. A basketball player is standing directly in front of the basket at the foul line, which is 15 ft from the basket. The player passes the ball to another player 8 ft directly to the right. This player takes the shot from that point. If the three-point line is 23 ft 9 inches from the basket, will this shot count for three points if it is made?

7. A kite flying on a 150-ft string is anchored to the ground. At one point the string is making a 70° angle with the ground. How high above the ground is the kite?

8. A surveyor uses a transit to determine the height of the Empire State Building in New York City. When pointed at the top of the antenna on the building, the angle of elevation given by the transit is approximately 80° from a point 260 ft away from the building. To the nearest foot, how tall is the Empire State Building including the antenna?

9. In testing the acoustics for a theater, the distance from the stage to the last row in the balcony has to be determined. If the angle of depression from that row to the stage is 28° and the vertical height above the stage is approximately 50 ft, how far, to the nearest foot, does the sound have to travel to reach someone sitting there?

10. An outdoor patio along the back of a house has to be constructed on an incline to allow rainwater to drain away from the house. To the nearest tenth of a degree, what is the angle at which the patio inclines, assuming that the length across the patio extends 12 ft from the house and that it is 3.7 in higher at the side adjoining the house than on the other side?

11. A 25-ft vertical pole is to be supported by two cable wires, each 30 ft long, on different sides of the pole. To the nearest foot, what is the distance between the points at which the cables are anchored to the ground? What is the angle, to the nearest degree, that these wires make with the ground?

12. Find the area, to the nearest square centimeter, of a regular pentagon whose perimeter is 100 cm.

13. On a baseball diamond, the distances between the bases is exactly 90 ft. The catcher has to be able to make an accurate throw from homeplate to second base. If the catcher throws the ball at 80 mph, how many seconds, to the nearest tenth, will it take the ball to reach second base?

14. Find the exact area of a parallelogram whose sides measure 10 in and 20 in and the acute angle formed by the sides is 60°.

Solutions to Additional Problems

1. Let x be the multiple to use with the ratio. Therefore, the angles, in descending size order, are $5x$, $3x$, and $3x - 40$.

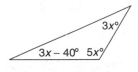

The fact that the sum of the measures of the angles of a triangle is 180° will give us the equation to solve.

$$5x + 3x + 3x - 40 = 180$$

$$11x = 220$$

$$x = 20$$

The largest angle is $5 \times 20 = 100°$.

Check: The middle angle is $3 \times 20 - 60°$, and $100° : 60°$ is indeed 5:3. The remaining angle is $60° - 40° = 20°$ and $100° + 60° + 20° = 180°.\checkmark$

2. The key point is that *the acute angles of a right triangle are complementary.* That is, their sum is 90°.

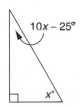

The relationship given provides the model

(Measure of smaller angle)^2 = 10 × measure of larger angle − 25

Let x be the measure of the smaller angle and $90 − x$ be the measure of the larger angle. Using these in the model, we have the quadratic equation

$$x^2 = 10(90 − x) − 25$$

(which needs to be simplified, and all terms must be brought to one side)

$$x^2 = 900 − 10x − 25$$
$$x^2 + 10x − 875 = 0$$

We seek factors of −875 whose difference is +10. A bit of trial and error will yield −25 and +35.
Therefore

$$(x − 25)(x + 35) = 0$$
$$x = 25 \quad \text{or} \quad −35$$

We can reject −35 since angles must have a positive measure. The smaller angle measures 25°, and the larger measures 65°.

Check: $25^2 = 625$ and $10 × 65 − 25 = 650 − 25 = 625.\checkmark$

3.   The leg of the trapezoid can be regarded as a *transversal* that, if extended, would cross the *parallel* sides.

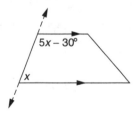

The two angles are, hence, interior angles on the same side of the transversal, and they are supplementary. The model is, therefore, the sum of the two angles $= 180°$.

$$\text{Let } x = \text{one angle}$$

$$5x - 30 = \text{the other angle}$$

$$x + 5x - 30 = 180$$

$$6x = 210$$

$$x = 35$$

The angles are $35°$ and $5 \times 35 - 30 = 175 - 30 = 145°$.

Check: We need to make sure that the angles are supplementary: $35° + 145° = 180°.\checkmark$

4. In an isosceles triangle, the *base* usually refers to the noncongruent side and the angles formed on this side are called the *base angles*. The third angle is called the *vertex angle* of the isosceles triangle. The diagram for this picture requires an isosceles triangle with a line segment that is clearly *not* an altitude, since the altitude to the base would also cut the vertex angle exactly in half.

Since we are dealing with ratios, we will assign a variable for the multiple as follows. Let x be the multiple. Therefore, $3x$ is the smaller angle formed by the segment and $5x$ is the larger one. Therefore, the vertex angle is $8x$. To avoid confusion, we can represent the base angle by a different variable. Let y be the measure of the base angle. We can use the sum of the angles of a triangle $= 180°$ to model the problem: $8x + y + y = 180$ or $8x + 2y = 180$. To solve a problem with two variables, we need a second equation involving both of them. This comes from the other piece of information, namely, the smaller angle formed by the segment is $20°$ less than the vertex angle. Therefore, we have the equation $3x = y - 20$ or $y = 3x + 20$. We can substitute for y in the first equation, giving us $8x + 2(3x + 20) = 180$. Simplifying and solving for x, we obtain

$$8x + 6x + 40 = 180$$
$$14x + 40 = 180$$
$$14x = 140$$
$$x = 10$$

Therefore, the angles formed by the segment are $3 \times 10 = 30°$ and $5 \times 10 = 50°$.

Check: The base angles would each have to be $20°$ more than the smaller angle, that is, $50°$. The sum of the angles of the isosceles triangle is $50° + 50° + 80° = 180°.\checkmark$

5. The phrase "vertically upward" indicates that the shaft is perpendicular to the horizontal (axis) and a right triangle is formed whose legs are the entrance and shaft and whose hypotenuse is the hillside.

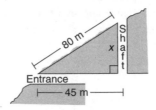

Therefore, we can apply *the Pythagorean Theorem*. Let x be the length of the vertical shaft. We have $x^2 + 45^2 = 80^2$ or $x^2 + 2025 = 6400$; Thus $x^2 = 4375$ and $x = \sqrt{4375} = 66.14$. To the nearest meter, the shaft is **66 m**.

6. In order for the shot to be worth three points, the distance to the basket has to be greater than or equal to 23 ft 9 in or 23.75 ft (since 9 in is three-fourths of a foot). The use of the word *directly* enables us to assume that we are dealing with perpendicular distances and, hence, we have the following diagram.

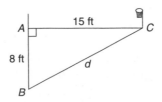

The first player is at A, the second player is at B, and the basket is at C. BC is the distance we are trying to determine, and it is the hypotenuse of a right triangle whose legs are AC and AB. We recognize these to be the legs of the 8, 15, 17 Pythagorean Triple, and the hypotenuse is 17, which is too short for a three-point shot. If you don't recognize the triple, you can use the Pythagorean Theorem. Let d be the distance, and we have $d^2 = 8^2 + 15^2 = 64 + 225 = 289$ and $d = \sqrt{289} = 17$ ft.

7. The diagram shows that we have a right triangle where the height of the kite is the side *opposite* the $70°$ angle and the string is the *hypotenuse*. Therefore, we make use of the sine ratio. Let h be the height of the kite. Then $\sin 70° = h/150$. From a calculator or table, we find $\sin 70° = 0.9397$. Therefore, $h = 150 \times 0.9397 = 140.96$. To the nearest foot, the kite is 141 ft above the ground.

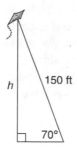

8. The diagram indicates that we have a right triangle. We see that 260 ft is the length of the side *adjacent* to the given acute angle and the height of the Empire State Building is the side *opposite* this angle. Therefore, we need to use the tangent ratio. Let x be the height of the building, and we have $\tan 80° = x/260$ or $x = 260 \times \tan 80° = 260 \times 5.6713 = 1474.5$ ft. To the nearest foot, the height is 1475 ft. (The actual height is 1472 feet. The difference is due to rounding the angle given by the transit.)

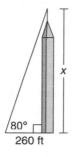

9. The angle of depression from the last row in the balcony to the stage is the same as the angle of elevation from the stage to that row. Therefore, we have a right triangle where the *hypotenuse* is the distance we wish to find and the vertical height is the side *opposite* the known acute angle.

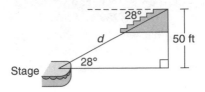

We can make use of the sine ratio. Let d be the distance, and we have $\sin 28° = 50/d$ or $d = 50 \div \sin 28° = 50 \div 0.4695 = 106.49$. The sound will have to travel approximately 107 ft.

10. To find the angle of incline, we will have to find a right triangle in which we know two of the sides. At the house side of the patio we have a vertical height of 3.7 in, and, since this is measured perpendicularly to the ground, we have a right triangle that extends to the other side of the patio, 12 ft or 144 in away. (We must work in the same units when taking ratios of sides.)

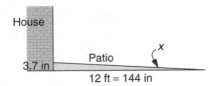

The diagram indicates that we have the sides *opposite* and *adjacent* to the angle of incline, which indicates that we need to use the tangent ratio. Let x be the measure of the angle of incline, and we have $\tan x° = 3.7/144 = 0.0257$. Using the inverse tangent, we have $x = 1.47°$ or, to the nearest tenth of a degree, $1.5°$.

11. The problem asks us to find two pieces of information. The model for this situation is easily seen to be an isosceles triangle where the cables are its congruent sides and the pole is its altitude. The altitude of a right triangle forms two congruent right triangles, and, since we know two of its sides, we can use the Pythagorean

Theorem to find the third. This side of the right triangle is one-half the distance between the cables on the ground.

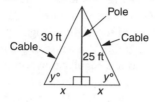

Let x represent the third side of the right triangle. The Pythagorean Theorem gives us

$$x^2 + 25^2 = 30^2$$
$$x^2 + 625 = 900$$
$$x^2 = 275$$
$$x = \sqrt{275} = 16.58$$

The distance between the cables is, therefore, 33.16 ft or, to the nearest foot, 33 ft. To find the angle, we could create a ratio with any two of the sides. However, we should use the given information rather than the length we computed, in case we computed incorrectly. The given sides represent the side *opposite* the angle and the *hypotenuse*. Therefore, we should use the sine ratio. Let y be the angle, and we have $\sin y° = 25/30 = 0.8333$. The inverse sine gives us $y = 56.4°$ or, to the nearest degree, 56°.

12. The diagram shows us that the interior of the pentagon can be divided into 10 congruent right triangles by drawing from the center all the radii and perpendicular segments to each side. Such a perpendicular segment is called an *apothem*. The area of the pentagon will be equal to 10 × the area of each right triangle.

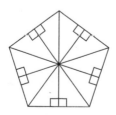

Since all sides are congruent, each side has a measure of 100 cm ÷ 5 = 20 cm. Since the radii form isosceles triangles, *the apothem bisects each side*. Therefore, we know that each leg of the right triangle is 10 cm. We need to find the length of the apothem in order to find the area of the right triangle. We can use trigonometry, if we can find the acute angle formed by a radius and a side. A basic fact about a regular polygon with n sides is that an exterior angle formed by extending a side is equal to $360/n°$. For a pentagon, this is $360 ÷ 5 = 72°$. The interior angle is its supplement, $180° - 72° = 108°$. The radius bisects this angle and, therefore, the acute angle in the right triangle is $54°$.

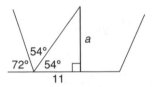

The apothem is the side *opposite* this angle and the "half side" of the pentagon is the side *adjacent* to this angle. Therefore, we can use the tangent ratio. Let a be the length of the apothem, and we have $\tan 54° = a/10$ or $a = 10 × \tan 54° = 10 × 1.3764 = 13.764$. The area of the right triangle is $\frac{1}{2} × 10 × 13.764 = 68.82$ cm^2. The area of the pentagon is therefore, $10 × 68.82 = 688.2$ cm^2 or, to the nearest square centimeter, **688 cm^2**.

13. The speed of the throw will come from the distance formula: Distance = rate × time; specifically, Time = distance ÷ rate. The rate is 80 mph, and we will have to convert units. The problem will therefore require several steps.

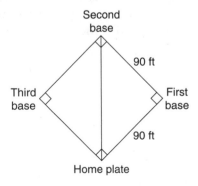

(*a*) Find the distance from home plate to second base. Home plate, first base, and second base form an isosceles right triangle with the distance we seek as the hypotenuse. Since this is a special 45–45–90 right triangle, we know that the hypotenuse is $\sqrt{2} \times 90$. Since we will be making conversions later, we should keep several decimal places along the way. $\sqrt{2}$ is approximately 1.4142, and our distance is approximately 127.278 ft.

(*b*) Convert the rate to feet per second. For this we can use the dimension scheme (see Chap.1)

$$\frac{85 \text{ mi}}{1 \text{ hour}} \times \frac{5280 \text{ ft}}{1 \text{ mi}} \times \frac{1 \text{ hour}}{60 \text{ minutes}} \times \frac{1 \text{ minute}}{60 \text{ seconds}}$$

$$= 124.667 \frac{\text{feet}}{\text{second}}$$

(*c*) Calculate the time.

Time $=$ distance $\div$ rate $= 127.278 \div 124.667 = 1.02$ seconds

or, to the nearest tenth, 1.0 second.

14. The area of a parallelogram is the product of one side (the base) and the altitude drawn to that side (the height). The word *exact* is an indicator that we will be able to express our answer either as a number that we did not have to round off or as a number involving radicals. In either case, we should be looking for a special right triangle.

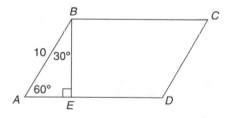

The diagram shows that we have a 30–60–90 right triangle and we know that the sides are in the ratio leg opposite the 30° angle : leg opposite the 60° angle: hypotenuse $= 1 : \sqrt{3} : 2$. We know that the hypotenuse AB is 10 in. The side opposite the 30° angle is, therefore, 5 in, and the side opposite the 60°, our altitude, is $5\sqrt{3}$ in. Therefore, the area of the parallelogram is 20 in $\times 5\sqrt{3}$ in $= 100\sqrt{3}$ in².

119

Table of Fundamental Geometric and Trigonometric Relationships

Definitions

Congruent	Equal in measure and form.
Similar	Proportional in measure, but identical in form.
Supplementary angles	Two angles whose sum is 180°.
Complementary angles	Two angles whose sum is 90°.
Adjacent angles	Two angles that share a common side and vertex.
Vertical angles	Two nonadjacent angles formed by intersecting lines.
Parallel lines	Two lines that are always the same distance apart and, therefore, will never intersect.
Regular polygon	A closed figure in which all sides and interior angles are congruent.
Bisector	A line or line segment that divides a side or angle into two congruent sides or angles.
Altitude	A line segment from a vertex of a triangle perpendicular to the opposite side. Its length is the *height* of the triangle when the opposite side is considered to be its *base*.
Median	A line segment from a vertex of a triangle drawn to the midpoint of the opposite side, therefore, bisecting the opposite side.

Relationships between Lines and Angles

1. Vertical angles formed by intersecting lines are congruent.
2. Adjacent angles forming a straight line are supplementary.
3. Adjacent angles forming a right angle are complementary.

Continued

4. Perpendicular lines form right angles.
5. A transversal that intersects a pair of parallel lines forms
   a. Congruent angles in corresponding positions on each parallel line
   b. Congruent angles interior to both parallel lines and on alternate sides of the transversal
   c. Supplementary angles interior to both parallel lines and on the same side of the transversal

Relationships Involving Triangles

6. The sum of any two sides of a triangle must be greater than the remaining side.
7. The sum of the angles of a triangle is 180°.
8. An exterior angle is equal to the sum of the two nonadjacent interior angles.
9. Two triangles are similar if they have the same three angles.
10. The ratios of corresponding sides of similar triangles are equal.
11. A median divides a triangle into two triangles of equal area.
12. In an isosceles triangle, two sides are congruent and their opposite angles are congruent.
13. In an isosceles triangle, the altitude drawn to the noncongruent side is also an angle bisector and a median.
14. In an equilateral triangle, all sides are congruent and each angle measures 60°.

Relationships Involving Right Triangles

15. The Pythagorean Theorem states that the square of the hypotenuse is equal to the sum of the squares of the legs.
16. The square of the length of an altitude drawn to the hypotenuse of a right triangle is equal to the product of the segments it creates on the hypotenuse.
17. The trigonometic ratios are
    a. *Sine* of an angle $= \frac{\text{length of opposite side}}{\text{length of hypotenuse}}$
    b. *Cosine* of an angle $= \frac{\text{length of adjacent side}}{\text{length of hypotenuse}}$
    c. *Tangent* of an angle $= \frac{\text{length of opposite side}}{\text{length of adjacent side}}$

Continued

Relationships Involving Quadrilaterals

18. The sum of the angles of any quadrilateral is 360°.
19. Opposite sides of a parallelogram are parallel and congruent.
20. Opposite angles of a parallelogram are congruent.
21. Consecutive angles of a parallelogram are supplementary.
22. The diagonals of a parallelogram bisect each other.
23. A rectangle, a rhombus, and a square are all parallelograms and have relationships 16 through 19.
24. A rectangle has four right angles, and its diagonals are congruent.
25. A rhombus has four congruent sides, and its diagonals are perpendicular.
26. A square is both a rectangle and a rhombus and has relationships 21 and 22.
27. A trapezoid has only one pair of parallel sides.

Relationships Involving Regular Polygons

28. The sum of the interior angles of an n-sided regular polygon is $180 \times (n - 2)$.
29. The measure of each exterior angle of an n-sided regular polygon is $360 \div n$, and the interior angle is its supplement.
30. The segments are drawn from the center of an n-sided regular polygon to the vertices from n congruent isosceles triangles.
31. The ratio of the areas of similar polygons is equal to the square of the ratio of corresponding sides or of the perimeters.

Relationships Involving Circles

32. All radii of the same circle or congruent circles are congruent.
33. The diameter is the longest line segment with endpoints on the circle.
34. A radius meets a tangent to a circle at a right angle.
35. Opposite angles of a quadrilateral inscribed in a circle are supplementary.

Chapter 5

Word Problems Involving Statistics, Counting, and Probability

Many word problems will ask you to analyze data and to make calculations that describe the data. These calculations include familiar statistical ideas such as mean, median, and mode and computing probabilities. A necessary ingredient in all of these is the ability to count objects accurately. In the problems that follow, you will see the most common ways of doing this and the variety of situations that appear most frequently when working with statistics, counting, and probability.

Mean, Median, and Mode

The words *mean* and *average* are synonyms. You have probably computed your test average sometime in your life and know that for a set of scores

$$\text{Average score} = \frac{\text{sum of scores}}{\text{number of scores}}$$

The average score is descriptive of the data in the following way: *If all the data were the same, they would all be equal to the mean or average score.* For example, if you were in a car traveling at an *average* speed of 55 mph, it would be as if you traveled exactly 55 mi every hour of your journey. In reality, you may have traveled different distances during each hour. However,

123

if you divided the total distance by the number of hours of your trip, your computation would be 55.

It is useful to remember that the sum of the scores = the average score × the number of scores.

Example 1

Dana's average in math is 82 after taking eight tests. To the nearest whole number, what will her new average be if she earned a 92 on her next test?

Solution 1

Using the relationship described above, the sum of her scores is equal to $82 \times 8 = 656$. If the new score is a 92, the sum will be $656 + 92 = 748$. Therefore, her new average will be $748 \div 9 = 83.1$ or 83. (The best way to check this answer is to calculate the average of eight scores of 82 and one score of 92.)

Example 2

Three numbers are in the ratio of 2:4:7. What is the largest of these numbers if their average is 52?

Solution 2

The relationship tells us that the sum of the three scores is $52 \times 3 = 156$. We need to apply the algebra of ratios (see Chap. 3). Let x be the multiple, and we are given the numbers represented by $2x$, $4x$, and $7x$. Therefore, $2x + 4x + 7x = 13x = 156$ and $x = 12$. The largest of the three numbers is $7 \times 12 = 84$.

Check: The other two numbers are $2 \times 12 = 24$ and $4 \times 12 = 48$. The average of the three numbers is $(24 + 48 + 84) \div 3 = 156 \div 3 = 52.\checkmark$

The *median* of a set of data is the score that occupies the middle position when the scores are arranged from smallest to largest. In the last example, 48 is the median of the three scores 24, 48, and 84. If the list has an even number of terms, the median is the average of the two in the middle of the list.

When the data list is large, it may be presented in the form of a table which indicates the score and the number of times it appears in the list. This number is called the *frequency* of the score. In Example 2, each score has a frequency of 1. In Example 1, if all the test scores were 82, its frequency would be 8. The *mode* of the data would be the score with the highest frequency, that is, the score that appears in the list most often.

Example 3

The Fahrenheit temperatures at 8 A.M. were recorded for each day during February in a Canadian city. The data are given in the table below. Find the mean, median, and mode temperatures.

Temperature	Frequency
$-4°$	4
$-2°$	6
$3°$	9
$4°$	4
$10°$	5

Solution 3

The mode is easily seen to be $3°$ since it has the highest frequency. The median would require listing all the scores in ascending order. By adding the frequencies, we know the total number of scores in the list. In this case, the list has 28 scores. Since this is an even number, we would average the scores in the 14th and 15th positions in the list.

$$-4, -4, -4, -4, -2, -2, -2, -2, -2, -2,$$
$$3, 3, 3, \boldsymbol{3}, \boldsymbol{3}, 3, 3, 3, 3, 4, 4, 4, 4, 5, 5, 5, 5, 5$$

The average of these two scores is the median, 3.

Rather than listing all the scores, an easier way to find the median is to count through the frequencies in the table. The $-4°$s and $-2°$s occupy the first 10 positions. The $3°$s occupy positions 11 to 19, which include the two middle positions.

The mean can also be computed from the table. The sum of all the scores is the sum of the products of each score and its frequency. For these data, the sum is $-4 \times 4 + -2 \times 6 + 3 \times 9 + 4 \times 4 + 10 \times 5 = -16 + -12 + 27 + 16 + 50 = 65$. If no specific directions are given, it is customary to compute the mean to one more decimal place than that of the original scores. Therefore, to the nearest tenth, the mean is $65 \div 28 = 2.3$.

Example 4

This example demonstrates how the median and the mean can be very different for a set of data.

Of James' seven tests this quarter in biology, the median score was 82 and the mean score was 76. If the average of his highest three tests is 90, what is the average of his lowest three scores?

Solution 4

Since we know the mean, 76, we can compute the sum of all seven scores; that is, the sum is equal to $76 \times 7 = 532$. In order to find the average of the three lowest scores, we can let x be their sum. Therefore, $x + 82 + 3 \times 90 = 532$. Multiplying and combining terms, we get $x + 352 = 532$ and, after subtracting 352 from both sides, we see that $x = 180$. The average of the three lowest scores is $180 \div 3 = 60$.

Check: We need to check that the sum of the seven scores we now know is $532 : 3 \times 60 + 82 + 3 \times 90 = 180 + 82 + 270 = 532.\checkmark$

Counting Objects

Word problems that ask for a count of objects would be very tedious if we had to list all the possibilities to know how many we have. In fact, we saw in the previous examples that counting doesn't necessarily have to be done this way if we know how many times a specific object appears in the collection. The following problems ask for a count of the possibilities of making selections from a set of objects.

Example 5

To get from class to outside the main entrance of the school, you can use any of three stairways to the first floor, either of two corridors to the main lobby, and either of two doors. How many different paths can you choose from?

Solution 5

This situation can be visualized by a tree diagram or flowchart that diagrams all the possibilities.

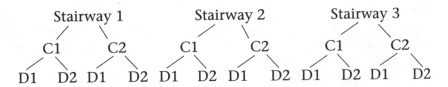

(where C1, C2 and D1, D2 represent corridors 1 and 2 and doors 1 and 2, respectively). There are clearly 12 possible ways from the classroom to the outside of the main entrance.

Note that the number of branches of the tree is based on the product $3 \times 2 \times 2 = 12$. This is the basis of the *Counting Principle*, which is a popular way to solve such problems. The Counting Principle states that *The total number of different selections consisting of picking one item from each of several different types is equal to the product of the possibilities for each item selected.*

Example 6

At the local fast-food restaurant you can choose from eight different sandwiches, four different side dishes, and five different beverages. How many different meals can you have consisting of one sandwich, one side order, and one beverage?

Solution 6

The Counting Principle leads us to believe that the number of different meals is equal to the product $8 \times 4 \times 5 = 160$. It's that simple! Drawing a tree diagram would take much longer.

One way to create a visual for counting problems such as these is to think of filling slots. For the problem described above, we would draw

————— × ————— × —————
Sandwich   side order   beverage

and fill the slots with the number of different choices for each item, namely, 8, 4, and 5. This will help with more complicated problems.

Example 7

Student ID numbers all have eight characters. The first two characters are different letters chosen from the letters A through F. The remaining six characters are digits from 1 to 9 which can be repeated. How many different ID numbers are possible to create?

Solution 7

The situation is the same as filling eight slots. The first slot is filled with a letter from a set of six, while the second is filled with a letter from a set of five since we cannot repeat the letter we already used. All the rest are filled with a number from a set of nine each since we can repeat digits.
The model is

$$\underset{\text{Letter}}{-6-} \times \underset{\text{letter}}{-5-} \times \underset{\text{number}}{-9-} \times \underset{\text{number}}{-9-} \times \underset{\text{number}}{-9-} \times \underset{\text{number}}{-9-} \times \underset{\text{number}}{-9-} \times \underset{\text{number}}{-9-}$$

or $6 \times 5 \times 9^6 = 15{,}943{,}230$. (A tree diagram would have been impossible to draw!)

Sometimes answers such as the previous one will seem incorrect. However, when considering all possibilities, such large numbers will turn up, and they cannot be checked by simple means. Therefore, it is important that you are sure of the way you reasoned through the problem.

Example 8

How many ways can four out of seven students be seated in a row of four chairs? How many ways can all seven be seated in a row of seven chairs?

Solution 8

Filling slots for the first situation, we have $\underline{7} \times \underline{6} \times \underline{5} \times \underline{4} = 840$. For the second situation we continue the product until we reach 1, $7 \times 6 \times 5 \times 4 \times 3 \times 2 \times 1$. This product is symbolized by 7! and is called "7 factorial": $7! = 5040$.

128

Your calculator may have a key for computing factorials, $n!$. You would enter the number and press the key for the computation. Another symbol sometimes used is $_nP_r$, where n is the number of objects in the set and r is the number of objects chosen to be arranged. The P stands for the term *permutation*, which is a synonym for *arrangement*. The answer to the first question would be symbolized by $_7P_4$ and the second answer by $_7P_7$, which is 7!.

Combinations

The previous problems were solved by considering the order in which selections were made. Some situations involve counting groups and are not concerned with a specific order. For these problems, we often make use of another symbol, $_nC_r$, where C stands for *combinations*. The formula is $_nC_r = {_nP_r}/r!$. This makes sense since there have to be less to count if we don't care in what order they appear. For example, the three-letter arrangements *abc, acb, bac, bca, cab,* and *cba* are *different permutations*, but the *same combination*.

Example 9

Mr. Fredricks wants to have a mock election in his social studies class. Students have to elect four people to serve as class officers. If nine students were nominated, how many possible different groups of four could the class elect?

Solution 9

If the problem asked for specific officer positions, we would be counting permutations. Since it does not, we will find the number of *combinations*. We use

$$_9C_4 = \frac{_9P_4}{4!} = \frac{9 \times 8 \times 7 \times 6}{4 \times 3 \times 2 \times 1}$$

Since our answer is going to be a whole number, we should try to reduce before multiplying and eliminate all the factors in the denominator. The fraction reduces nicely to 126, our answer.

Sometimes you won't be told explicitly that you are look-ing for combinations. You will often be able to find a way to restate the problem that makes it obvious.

Example 10

How many different diagonals can be drawn in a regular oc-tagon (an eight-sided figure with equal sides)?

Solution 10

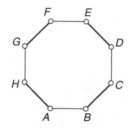

If you don't recognize the use of combinations, you can label the vertices *A* through *H* and start listing the diagonals sys-tematically: *AC, AD, AE, AF, AG, BD, BE, BF, BG, BH, CE, CF, CG, CH,* (*CA* is already in the list), *DF, DG, DH,* (*DA* and *DB* are already in the list), *EG, EH,* (the rest with *E* are already in the list), *FH* (the rest with *F* are already in the list). We count 20 diagonals.

If you think of a diagonal as being a connection between two vertices, the problem is really one of finding all the differ-ent pairs that can be formed from the eight vertices. That is

$$_8C_2 = \frac{_8P_2}{2!} = \frac{8 \times 7}{2 \times 1} = 28$$

However, eight of these pairs are not diagonals, but sides of the octagon. Therefore, the answer is $28 - 8 = 20$.

Probability

Most questions about probability are given as word problems with descriptions of practical situations. Several words are used to help clarify the situations so that a model for probability problems can be formed. The situation usually describes making selections and is referred to as the *experiment*. When considering all the ways in which the situation can unfold, there may be several *outcomes* that are considered to be *successful* in achieving the desired *event*. The probability is a simple ratio.

$$\text{Probability of an event} = \frac{\text{number of successful outcomes}}{\text{total number of outcomes}}$$

This will always be a number ranging from 0 to 1, since the numerator can never be larger than the denominator.

Once again, we find ourselves concerned with counting. Therefore, we should expect to make use of the Counting Principle and, if necessary, the formulas for permutations and combinations.

Example I I

A bag of candy contains candies of different fruit flavors. Four candies are strawberry-, eight are lemon-, and three are cherry-flavored. If two candies are picked at random without replacement, what is the probability that they both will have the same flavor?

Solution I I

To solve this problem, we have to identify the *experiment*, an *outcome*, and what constitutes a *successful* outcome.

Experiment: Picking two candies.
An outcome: The two flavors picked.
A successful outcome: The flavors are the same.

We now need to count the total number of outcomes. For this, we can use the Counting Principle. Since we are not replacing the first candy and there are 15 candies in the bag, there are $15 \times 14 = 210$ different ways to pick 2 of them.

To find the number of successful outcomes, we have to add the number of ways we can pick two candies of each flavor. There are $4 \times 3 = 12$ ways to pick strawberry candies,

$8 \times 7 = 56$ ways to pick lemon candies, and $3 \times 2 = 6$ ways to pick cherry candies. The number of successful outcomes is, therefore, $12 + 56 + 6 = 74$.

The probability of picking two candies of the same flavor is 74/210 or, when reduced, 37/105. (Sometimes you may be asked to express the answer as a decimal or percent. This may require rounding. To the nearest thousandth, the answer is 0.352; to the nearest percent, the answer is 35%.)

Many probability problems involve familiar objects such as a deck of 52 playing cards, a pair of dice, a circular spinner divided into colored regions, or a two-sided coin. Some points to remember are

- A regular deck of 52 playing cards consists of four suits, each having and ace, the numbers 2 through 10, a jack, a queen, and a king. Of these 13 labels, the jack, queen, and king are referred to as picture cards. Jokers are not part of the 52 cards.
- A *single die* is a cube with each face representing a number from 1 to 6. A *fair die* is one in which there is an equal chance of landing on either side. The top side is considered to be the result of the toss of the die. When a pair of dice is rolled, the sum of the two top sides will be any number from 2 to 12, but there are different probabilities for each of these sums.
- A *circular spinner* is divided into regions by radii drawn from the center. These regions may not necessarily be of equal area. The probability of landing in any region is the ratio of the area of the region to the area of the entire circular region and is directly proportional to the angle formed by the radii that border the region. (The ratio of areas is used for probabilities in any geometric setting.)
- A *fair coin* is one with which there is an equal probability of landing on either side. The sides are usually distinguished by "heads" and "tails," although it could have any kind of markings on either side. The probability of getting either heads or tails when tossing a fair coin is $1/2$. Sometimes we are told that the coin is *not* fair and there is a *bias* for one side.

132

Example 12

A card is drawn from an ordinary deck of 52 playing cards and replaced in the deck. A second card is then drawn. What is the probability that both are picture cards?

Solution 12

Identifying the components of the problem, we have

The experiment: Pick two cards.
An outcome: The types of cards picked.
A successful outcome: The two cards are picture cards (jacks, queens, or kings not necessarily the same).

Since we are replacing the first card, there will still be 52 possibilities for the second pick. Therefore, there are $52 \times 52 = 2704$ different outcomes. In the deck, there are 12 picture cards and the number of successful outcomes is $12 \times 12 = 144$. Therefore, the probability is $144/2704$.

We can also solve the previous problem by a different use of the Counting Principle, which deals with probabilities. Specifically, it states: *The probability of an event is equal to the product of the probabilities of being successful at each stage of the experiment.* The probability of picking a picture card at each stage is $12/52$ or $3/13$. Therefore, the probability of picking two picture cards is found simply by $3/13 \times 3/13 = 9/169$. Note how this helps with expressing the answer in reduced form.

Example 13

A fair coin is tossed 5 times. What is the probability of getting five heads?

Solution 13

The probability of getting a head on one toss is $1/2$. Using the modified Counting Principle, our answer is simply $1/2 \times 1/2 \times 1/2 \times 1/2 \times 1/2 = (1/2)^5 = 1^5/2^5 = 1/32$.

Problems involving probabilities sometimes involve data from real-world situations. For these problems, we think of the probability as the fraction or percentage that represents the number of times we are successful. The probability is used as the basis of a *prediction* or *expectation*. There is no real difference between this and the previous computations for probability, except that we usually have exact counts to work with.

Example 14

The monthly rainfall for a Midwestern (U.S.) city for the past year is given in the table below. How many months would you expect rainfall between 2 and 3 in over a 5-year period?

Month	Jan	Feb	Mar	Apr	May	Jun	Jul	Aug	Sep	Oct	Nov	Dec
Inches	2.9	2.5	3.3	4.2	4.8	4.3	4.1	3.3	3.5	3.1	3.3	2.9

Solution 14

The solution is quite simple. Since 3 months out of 12 meet the condition, the probability of between 2 and 3 in of rain in any month is 3/12 or 25%. Therefore, we expect that over a 5-year period which is 60 months, there will be 25% of 60 or 15 months in which the rainfall is between 2 and 3 in.

Summary

1. Be sure to identify all situations carefully and determine exactly what it is you are counting.
2. Determine whether you are or are not concerned with the specific order; that is, determine whether you will use permutations or combinations.
3. Describe the *experiment, outcomes,* and *successful outcomes* in order to understand the situation.

Additional Problems

1. So far this season, the average spectator attendance at the nine high school basketball games is 135. Each ticket costs $1.50, and the school expects the total revenue for the year to be $2610. What is the expected average attendance for the remaining three playoff games?
2. Six people were raising money for a charity. Jordan and Danny averaged $120, Michelle and Lindsey averaged $105, and Mark and Evan averaged $92. What was the average amount for all six?

3. Five consecutive odd integers are such that the sum of the first three is 60 more than the sum of the last. What is the mean and median of these numbers?

4. Theresa ordered a combo pizza with three different toppings. If there are six toppings to choose from and one of her toppings is mushrooms, how many possible different combo pizzas did she have to choose from?

5. Windstar Airlines flies from Newark to three different cities. From each of these cities, there are six different airlines that will take you nonstop to Dallas. On the return trip, if you do not take the same route, how many different itineraries can you plan starting and returning with Windstar at Newark?

6. Janice is completely unprepared for the quiz on the novel that was read in English class. The quiz consists of 10 multiple-choice questions with four choices per question. She decides to guess at each question. How many different answer sheets could Janice possibly produce?

7. There are 12 runners in a race. How many different outcomes are there for first, second, and third place?

8. The Senate Committee on Armed Services has 11 Republicans and 9 Democrats. They decide to form a subcommittee consisting of 4 Republicans and 3 Democrats. How many different subcommittees are possible?

9. During the first 2 years of college at the State University, a student has to take 2 math courses, a science course, and 3 social science courses. How many possible different combinations of courses can a student take if the school offers 6 math courses, 4 science courses, and 10 social science courses for freshmen and sophomores?

10. During the junior year of high school, the English classes must read 3 of 5 novels, 4 of 10 short stories, 1 of 3 Shakespearian plays, and 5 of 9 poems. What is the likelihood that two different English classes will read the same selections?

11. A biased coin is weighted so that the probability of landing heads is 2 in 3. If three coins are tossed simultaneously many times, what percent of the time will they all land tails?

12. In the game of poker, a "flush" is a 5-card hand consisting of cards all of the same suit. From a regular deck of 52 playing cards, what is the probability of picking 5 cards and winding up with a flush?

13. Stacy had to get dressed quickly for school one morning. She reached into her closet for pants and a blouse. She has three pairs

of blue pants, two pairs of black pants, and two pairs of brown pants. She also has three white blouses, four blue blouses, and one red blouse. What is the probability that she picked blue pants and a blue blouse?

14. The test scores of a class are given in the table below. While going over the test with the class, the teacher randomly picks two test papers out of the pile simultaneously to use as models. What is the probability that these papers will have scores greater than the average of the class?

Scores	Frequencies
55	3
62	3
65	5
78	1
83	3
88	9
93	4
100	2

15. A circular spinner is divided into four regions of different colors by four radii. If the ratios of the distinct angles formed by the radii are 1:3:5:11, what is the probability of the spinner landing in one of the two smaller regions?

16. On a popular TV game show, the contestant has to spin a wheel. The wheel is divided into 20 regions. All these regions are equal in area except for two which award the player the grand prize. If these regions are each one-fifth the size of the others, to the nearest hundredth, what is the probability that the contestant will win the grand prize?

17. On a circular dartboard whose diameter is 18 in, the bull's-eye in the center has a diameter of 6 in. What is the probability of a thrown dart landing in the bull's-eye?

18. On his way out the door to go to school in the morning, Brandon's mother tells him to take three different pieces of fruit from the bin. If there are five bananas, three oranges, and two pears, what is the probability that he will take two bananas and a pear?

19. When emptying her pocket one night, Gayle found that she had eight nickels, five dimes, five quarters, and two pennies. The next

day she put these coins in her pocket. At lunchtime, she reached in and took out two coins. What are the chances that she took out less than 35¢?

20. In an envelope are 10 straws each measuring a different whole-numbered length from 2 to 11 in. When taking three different straws out of the envelope without looking, what is the probability that the three straws could form a triangle?

Solutions to Additional Problems

1. We need to work backward from the expected revenue and compute the expected total attendance for the 12 games. At $1.50 per ticket, the expected total attendance is $2610 ÷ $1.50 per ticket = 1740 tickets sold. Using the fact that the sum of the attendance figures equals the average attendance times the number of games, we know that the total attendance of the nine games played so far is $9 \times 135 = 1215$. Therefore, the school needs to sell $1740 - 1215 = 525$ more tickets. The average attendance for the remaining three games would be $525 ÷ 3 = 175$.

2. We need to compute the total amount of money raised by all six people. Jordan and Danny raised $120 \times 2 = $240, Michelle and Lindsey raised $105 \times 2 = $210, and Mark and Evan raised $92 \times 2 = $184. The total is $240 + $210 + $184 = $634. Therefore, the average for each person is $634 ÷ 6 = $105.67.

3. We need to use some algebra for this. Let x be the first integer. The others are $x + 2$, $x + 4$, $x + 6$, and $x + 8$. The sum of the first three is $x + x + 2 + x + 4 = 3x + 6$. Therefore, our equation is

$$3x + 6 = x + 8 + 60 \rightarrow \quad 3x + 6 = x + 68 \rightarrow$$
$$2x = 62 \rightarrow \quad x = 31$$

If the first odd integer is 31, the others are 33, 35, 37, and 39. The mean and median are the same, namely 35.

4. The Counting Principle applies to this problem with three slots to fill. Since we know that one of Theresa's toppings is mushrooms, the first slot has only one choice. The remaining toppings are different. Therefore, the second slot has five choices and the third has four. There are $1 \times 5 \times 4 = 20$ different combo pizzas that she can create.

5. There are four slots to fill in this problem using the Counting Principle.

$$\underline{\qquad} \times \underline{\qquad} \times \underline{\qquad} \times \underline{\qquad}$$

| From Newark with Windstar | airline to Dallas | airline from Dallas | to Newark with Windstar |

Since you cannot take the same Windstar routes or the same airline to and from Dallas, the number of different itineraries is the product $3 \times 6 \times 5 \times 2 = 180$.

6. The Counting Principle is applied with 10 slots, each representing one question with four possible choices for each question. Therefore, the number of different answer sheets is the very large number $4^{10} = 1,048,576$. (Note that the probability that Janice will answer all the questions correctly is $1/1,048,576$ or approximately 0.000000095, a very small chance!)

7. This is a permutation problem since we are interested in the arrangement of the three people who finish in the first three positions. The answer is $_{12}P_3 = 12 \times 11 \times 10 = 1320$ possible outcomes for the race.

8. The solution to this problem involves both the Counting Principle and combinations. That is, each possible group of Republicans can be joined with any of the possible groups of Democrats. Therefore, the number of different subcommittees will be the product of the two numbers of ways we can pick groups of each. The number of groups of Republicans is

$$_{11}C_4 = \frac{_{11}P_4}{4!} = \frac{11 \times 10 \times 9 \times 8}{4 \times 3 \times 2 \times 1} = 330$$

The number of groups of Democrats is

$$_9C_3 = \frac{_9P_3}{3!} = \frac{9 \times 8 \times 7}{3 \times 2 \times 1} = 84$$

The number of different subcommittees is $330 \times 84 = 27,720$.

9. The Counting Principle can be applied with the slots as

$$\underline{\qquad} \times \underline{\qquad} \times \underline{\qquad}$$

| Math courses | Science courses | Social science courses |

There are

$$_6C_2 = \frac{_6P_2}{2!} = \frac{6 \times 5}{2 \times 1} = 15$$

different ways to choose math courses. There are

$$_4C_1 = \frac{_4P_1}{1!} = \frac{4}{1} = 4$$

different ways to choose a science course. There are

$$_{10}C_3 = \frac{_{10}P_3}{3!} = \frac{10 \times 9 \times 8}{3 \times 2 \times 1} = 120$$

different ways to choose social science courses. Therefore, there are $15 \times 4 \times 120 = 7200$ different combinations of these courses that a freshman or sophomore can take.

10. The word *likelihood* is a synonym for *probability*. The problem needs to be restated in a way that identifies the event. Two classes reading the same selections is the same as already knowing the selection that one class has made and asking, "What is the probability of that being the selection for another class?" The probability is, therefore

$$\frac{1}{\text{Number of possible different selections}}$$

There are $_5C_3 = 10$ different groups of 3 novels, $_{10}C_4 = 210$ different groups of short stories, 3 different Shakespearian plays, and $_9C_5 = 126$ different groups of poems. Therefore, using the Counting Principle, there are $10 \times 210 \times 3 \times 126 = 793,800$ different selections to choose from. The probability is $1 \div 793,800 = .000001259$ or *there is almost no likelihood that this will happen!*

11. A key fact about probability is that *the sum of all possible outcomes is 1*. In other words, if *any* outcome is considered to be successful, the numerator and denominator of the probability fraction must be equal. Since the only outcomes for tossing a coin are heads and tails and we are given the probability that landing heads is 2/3, the probability of landing tails is 1/3. We use the modified Counting Principle for probabilities and find the probability of three tails on one toss of the three coins to be $1/3 \times 1/3 \times 1/3 = 1/27$ or 0.037. Therefore, we can expect this to occur 3.7% of the time.

12. A "flush" is a group of cards of the same suit in no specific arrangement. Therefore, for any one suit, the number of different

flushes is $_{13}C_5 = 1287$. Since there are four different suits, there are $4 \times 1287 = 5148$ different flushes. The total number of different five-card hands is $_{52}C_5 = 2{,}598{,}960$. Therefore, the probability is $5148 \div 2{,}598{,}960 = 0.00198$. (Since this number is nearly 0, there is an extremely small chance that this will occur!)

13. Identifying the components of the problem, we have *the experiment*—picking pants and picking a blouse (each pick represents a different stage of the experiment); *an outcome*—picking any of seven pairs of pants and any of eight blouses; and *a successful outcome*—picking any of three pairs of blue pants and any of four blue blouses. We can use the modified *Counting Principle for Probabilities*, which states that the probability of a successful outcome is *the probability of picking blue pants* $\times$ *the probability of picking a blue blouse*. Using combinations, we have the probability of picking blue pants $= 3/7$ and the probability of picking a blue blouse $= 4/8 = 1/2$. The answer is, therefore, $3/7 \times 1/2 = 3/14$.

14. Identifying the components of the problem, we have *the experiment*—picking two papers without replacement (each pick represents a different stage of the experiment); *an outcome*—picking any pair of papers; and *a successful outcome*—picking a pair greater than the average. To find the average score, we have to get the sum of the scores by multiplying each score by the number of times it appears in the set of 30 and then add the products. This is easiest to do by adding a third column to the table which will contain the products and adding a row at the bottom that will contain the sums of the second and third columns.

Score	Frequencies	Products
55	3	165
62	3	186
65	5	325
78	1	78
83	3	249
88	9	792
93	4	372
100	2	200
Sums	*30*	*2367*

The average is 2367 ÷ 30 = 78.9. The scores of 83, 88, 93, and 100 are all greater than the average. The frequencies indicate that there are 18 of these scores. Therefore, the probability of the first paper having a score greater than the average is 18/30. Since the second paper is pulled out at the same time as the first, we lose one possibility from those greater than the average and from the total pile. Therefore, its probability would be 17/29. The modified Counting Principle for Probabilities states that the probability of both papers having scores greater than the average = 18/30× 17/29 = 0.35, rounded.

Note that this answer is equivalent to

$$\frac{_{18}C_2}{_{30}C_2} = \frac{(18 \times 17)/(2 \times 1)}{(30 \times 29)/(2 \times 1)} = \frac{18 \times 17}{2 \times 1} \times \frac{2 \times 1}{30 \times 29} = \frac{18 \times 17}{30 \times 29}$$

which indicates that the problem could have been solved by finding the ratio of (*a*) the number of ways to select 2 from the 18 scores that are greater than the average to (*b*) the number of ways of selecting 2 scores from the entire set of 30.

15. The probability of a spinner landing in a region is the ratio of the area of the region to the area of the circle. The ratio of the areas of the regions is directly proportional to the ratio of these central angles. While it is not necessary to compute the angles, it is helpful in understanding the situation. Using the algebra of ratios, let *x* be the multiple and let the angles be *x*, 3*x*, 5*x*, and 11*x*.

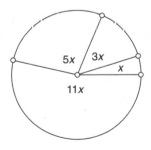

Their sum is 20*x*. Since the entire rotation about the center of the circle is 360°, we have 20*x* = 360 and *x* = 18. Therefore, the two smaller angles are 18° and 54°. Together this is 72/360 of the rotation. The probability of landing in either of the two smaller regions is, therefore, $72/360 = 2/10$ or 0.2.

16. In this problem, we are given information about the areas of the region. The probability of a spinner landing in a region is the ratio of the area of the region to the area of the circle. We can use algebra to determine this ratio. Since the smaller regions are one-fifth the size of the others, we let x be the area of each of the smaller regions and $5x$ be the area of each of the other 18 regions. The total area of the regions we wish to land in is $2x$, and the total area of the wheel is $5x \times 18 + 2x = 92x$. Therefore, the probability of landing in either of the smaller regions is $2x/92x = 1/46$ or, to the nearest hundredth, 0.02.

17. This problem is similar to Probs. 15 and 16 as it involves finding the ratio of areas. (This, in fact, will be true of any problem involving interior regions of a geometric shape. Such problems are referred to as *problems of geometric probability*.) The probability of the dart landing in the bull's-eye is the ratio of the area of the bull's-eye to the area of the dartboard. The "bull's-eye" is a circle whose center is that of the larger circular dartboard.

Since we know the diameters of each circle, 6 and 18 in, we know that the radii are 3 and 9 in. The area of the bull's-eye is 9π and the area of the entire dartboard is 81π. Therefore, the probability is $9\pi \div 81\pi = 1/9$. Note that this is the square of the ratios of the radii: $(3/9)^2 = (1/3)^2 = 1/9$. This will always be true about the ratio of areas of two circles.

18. This problem can be solved using the same idea as in Prob. 8, using combinations. The probability is the ratio of the number of ways Brandon can pick 2 bananas and 1 pear to the number of ways he can pick any 3 fruits from the collection of 10 fruits. The

answer is

$$\frac{_5C_2 \times {_2}C_1}{_{10}C_3} = \frac{10 \times 2}{120} = \frac{20}{120} = \frac{1}{6}$$

19. When all else fails, a probability problem can be solved by listing all possible outcomes. In this case a table whose column and row headings indicating the different coins can be used. Each cell contains the total value of the two coins. Note that the diagonal contains no values since we cannot use the same coin twice.

	p	p	n	n	n	n	n	n	n	n	d	d	d	d	d	q	q	q	q	q
p	X	2	6	6	6	6	6	6	6	6	11	11	11	11	11	26	26	26	26	26
p	2	X	6	6	6	6	6	6	6	6	11	11	11	11	11	26	26	26	26	26
n	6	6	X	10	10	10	10	10	10	10	15	15	15	15	15	30	30	30	30	30
n	6	6	10	X	10	10	10	10	10	10	15	15	15	15	15	30	30	30	30	30
n	6	6	10	10	X	10	10	10	10	10	15	15	15	15	15	30	30	30	30	30
n	6	6	10	10	10	X	10	10	10	10	15	15	15	15	15	30	30	30	30	30
n	6	6	10	10	10	10	X	10	10	10	15	15	15	15	15	30	30	30	30	30
n	6	6	10	10	10	10	10	X	10	10	15	15	15	15	15	30	30	30	30	30
n	6	6	10	10	10	10	10	10	X	10	15	15	15	15	15	30	30	30	30	30
n	6	6	10	10	10	10	10	10	10	X	15	15	15	15	15	30	30	30	30	30
d	11	11	15	15	15	15	15	15	15	15	X	20	20	20	20	35	35	35	35	35
d	11	11	15	15	15	15	15	15	15	15	20	X	20	20	20	35	35	35	35	35
d	11	11	15	15	15	15	15	15	15	15	20	20	X	20	20	35	35	35	35	35
d	11	11	15	15	15	15	15	15	15	15	20	20	20	X	20	35	35	35	35	35
d	11	11	15	15	15	15	15	15	15	15	20	20	20	20	X	35	35	35	35	35
q	26	26	30	30	30	30	30	30	30	30	35	35	35	35	35	X	50	50	50	50
q	26	26	30	30	30	30	30	30	30	30	35	35	35	35	35	50	X	50	50	50
q	26	26	30	30	30	30	30	30	30	30	35	35	35	35	35	50	50	X	50	50
q	26	26	30	30	30	30	30	30	30	30	35	35	35	35	35	50	50	50	X	50
q	26	26	30	30	30	30	30	30	30	30	35	35	35	35	35	50	50	50	50	X

There are $20 \times 20 - 20 = 380$ filled cells (note that this is 20×19). Of these, 70 are not successful; therefore, the probability is $310/380 = 31/38$, or the chance that Gayle took out less than 35¢ is approximately 0.82. Identifying the components for this problem, we have *the experiment*—picking two coins; *an outcome*—pick any two coins totaling any amount; and *a successful outcome*—picking two coins whose total value is less than 35¢. The total number of outcomes is

$$20C_2 = \frac{20P_2}{2!} = \frac{20 \times 19}{2 \times 1} = 190$$

which will be the denominator of the probability ratio. (Note that we are not considering the order in which we pick the coins; that is why this is half of the number we saw in that table.) It should be clear from the problem that there are considerably fewer ways to pick two coins unsuccessfully than successfully. We can find the number of ways to do this and subtract. Any pick of two quarters is unsuccessful and there are

$$5C_2 = \frac{5P_2}{2!} = \frac{5 \times 4}{2 \times 1} = 10$$

ways to do this. Any pick of a quarter and a dime will be unsuccessful, and there are $5 \times 5 = 25$ ways of doing this. Therefore, our probability is $(190 - 35)/190 = 155/190 = 31/38$.

20. *The experiment*—picking three different straws; *an outcome*—three different lengths; and *a successful outcome*—the three lengths can represent the sides of a triangle. At first thought, you might say that the probability should be 1, thinking that you can form a triangle from any three lengths. This, however, is *not* true! A basic fact about triangles is that *the sum of any two sides must be greater than the third side*. Since there is more to consider than just counting the number of different groups of 3, it would be reasonable to find a *systematic* way to list the combinations and pick out the ones that work. Note that there are $_{10}C_3 = 120$ possible different triplets that can result and we should list the possibilities in a way that makes it easy to see that we have considered all possibilities. Therefore, we can start a table and include in the cells only those possible third sides which were not considered previously. For example, we can start with supposing that our first two picks are 2 and 3. We can have a side of 4, but anything higher would not be less than $2 + 3$. When moving down the column, we go to 2,4. A possible third side is 3, but we already listed the triplet 2,3,4 in the row above. The only other remaining possibility is 5. We move along through the 2s until 2,11, which would add no additional successful triplets, and then to the next column, where we consider the pairs, including a 3.

First two	Possible third	First two	Possible third	First two	Possible third	First two	Possible third	First two	Possible third	First two	Possible third	First two	Possible third	First two	Possible third	First two	Possible third
2,3	4																
2,4	5	3,4	5,6														
2,5	6	3,5	6,7	4,5	6,7,8												
2,6	7	3,6	7,8	4,6	7,8,9	5,6	7,8,9,11										
2,7	8	3,7	8,9	4,7	8,9,10	5,7	8,9,10,11	6,7	8,9,10,11								
2,8	9	3,8	9,10	4,8	9,10,11	5,8	9,10,11	6,8	9,10,11	7,8	9,10,11						
2,9	10	3,9	10,11	4,9	10,11	5,9	10,11	6,9	10,11	7,9	10,11	8,9	10,11				
2,10	11	3,10	11	4,10	11	5,10	11	6,10	11	7,10	11	8,10	11	9,10	11		
2,11	X	3,11	X	4,11	X	5,11	X	6,11	X	7,11	X	8,11	X	9,11	X	10,11	X

We can now count the number of entries in the "Possible third" column 70. Therefore, the probability of picking three straws that could form a triangle is $70/120 = 7/12$ or approximately 0.58.

Chapter 6

Miscellaneous Problem Drill

This collection of problems will give you an opportunity to use the skills, concepts, and strategies discussed in the previous chapters. If you need help, you can look back at the solutions to the examples and additional problems in the chapter mentioned before each set of problems. The answers are given at the end. Good luck!

Problems Similar to Those in Chap. I

1. Jim and Donna would like to put a fancy molding around their den. Jim likes the molding that costs $2.50/ft, while Donna prefers the molding that costs $1.75/ft. If the rectangular den is 15 ft × 12 ft, how much more would Jim be willing to pay for molding than Donna would?

2. The distance from homeplate to first base at the local schoolyard is 60 ft, and the entire diamond (which is really a square) is covered with grass except for the pitcher's circle, which is dirt. If the grass covers approximately 3550 ft^2, what is the diameter of the pitcher's circle to the nearest foot?

3. Laura walked to the store from her house in 20 minutes and then from the store to her friend's house, which was 1 mi away, in 15 minutes. If she walked at the same rate on both trips, how far is her house from the store?

4. When Danny first arrived in Italy on his trip, he weighed himself and found that he weighed 63 kg. When he returned home from his trip, he weighed himself again and found that he then weighed 142 lb. How many pounds did Danny gain or lose while in Italy?

5. A chef had to prepare soup for 50 people and used the entire capacity of the pot. If each soup bowl holds ¾ cup of soup, how many second servings of soup could the chef serve, if the pot has a 3-gal capacity?

6. Last year Gabriella invested $3500 in a mutual fund that had an annual return of 12 percent. This year she invested all the principal and interest from that fund into another fund that earned only 6 percent. How much money does she have now?

7. In November, Josh wanted to buy a football helmet that cost $49.99. The salesperson told him that after the football season in February the helmet would sell for 15 percent less. However, there was another helmet in the store that sold for $42.99. What should Josh do in order to pay the least amount for a helmet?

Problems Similar to Those in Chap. 2

8. The larger of two numbers is 8 less than 3 times the smaller. Given that the sum of the numbers is 88, find the larger number.

9. Find four consecutive odd integers whose sum is 96.

10. The sum of the squares of the first and last of three consecutive positive integers is 164. Find the three integers.

11. The denominator of a fraction is 3 more than the numerator. If 3 is added to both the numerator and the denominator, the new fraction has a value of 0.7. Find the original fraction.

12. The tens digit of a two-digit number is 3 less than the units digit. The number is 3 more than 4 times the sum of the digits. Find the number.

13. Today Allen is 8 years older than Harvey. In 6 years he will be twice as old as Harvey will be then. How old is Harvey now?

14. Jeff and Steve left their homes, which are 21 mi apart, at the same time and bicycled along the same straight road toward each other. If Jeff's speed was 4 mph faster than Steve's, how fast was Jeff traveling if they met each other in 45 minutes?

15. A nuclear reaction projects an atomic particle along a straight line at 900 m per second (m/second). A second particle is projected 5 seconds later along the same path at a speed of 1200 m/second. In how many seconds into its trip will the second particle pass the first?

16. The prices of two kinds of coffee are $1.75/lb and $3.40/lb. The owner of the store creates a 50-lb mixture of the two that sells for

$2.25. Assuming that all 50 lb are sold, how much, to the nearest pound, of the cheaper coffee should she use in the mixture to make sure that she makes the same money as if she sold the coffees separately?

17. Amy has twice as many nickels as quarters and two more dimes than nickels. If she has $1.85 in change, how many dimes does she have?

18. Carole is asked to do a job that ordinarily takes her 45 minutes to complete. Melissa usually does the same job in 30 minutes. If they work together, how long will it take for the job to be completed?

19. If two pumps are used to fill a tank at the same time, the tank is filled in 3 hours. The faster pump fills the tank by itself in 5 hours. This morning both pumps were turned on at the same time and left to fill the tank. After $1\frac{1}{2}$ hours, the slower pump stopped working. How much time did it take for the other pump to complete filling the tank?

20. Phyllis invested part of her $10,000 inheritance in a fixed annuity earning 7 percent annually. She invested the rest of her money in various stocks that were expected to earn 8 percent, but actually earned 12 percent over the year. How much did she invest in both situations if her earnings for the year were $1050?

Problems Similar to Those in Chap. 3

21. A school district boasts that there are three teachers for every 70 students. If the total number of teachers and students in the school is 4015, how many students are in the school?

22. Separate 144 into two parts so that the ratio of the two numbers is 5:7.

23. The ratio of two positive numbers is 4:1, and the sum of their squares is 153. Find the numbers.

24. The ratio of Gayle's age to Danielle's age is 3:2. In 10 years, Gayle's age will be 22 less than twice Danielle's age. How old is Gayle now?

25. Lindsey, Kristen, and Nicole ran for student body president. The ratio of votes received respectively were 2:5:6. If 2600 votes were cast, by how many votes did Nicole win the election?

26. A recipe in a cookbook calls for $\frac{1}{3}$ cup of sugar and 3 cups of flour to make a cake that serves eight people. Joan was expecting more than eight people and used 10 cups of flour. How many cups of sugar did she have to use?

27. In June the final exam in math given to all ninth graders is graded by two teacher committees. Committee A usually grades 30 more papers in an hour than committee B. In the time it took for all the papers to be graded, committee A graded 350 papers while committee B graded 280 papers. On the basis of these results, how many papers can each committee grade in one hour, and how many hours did it take for all the papers to be graded?

28. The chemist knows that the pressure of a gas within a container varies directly with the temperature in the container. If the pressure is measured at 35 lb/in^2 when the temperature is 18°C, how much higher, to the nearest tenth of a degree, must the temperature be raised so that the pressure reaches 40 lb/in^2?

29. In the manufacturing process of a cleaning fluid, 180 kg of a dry base is added to 500 L of water. The mixture is heated so that it becomes a 40 percent basic solution by weight. How much water must be eliminated in this process to achieve the proper mixture?

Problems Similar to Those in Chap. 4

30. In a triangle, the measure of the largest angle is 16° more than twice the measure of smallest angle and the third angle is 20° more than the smallest. What are the measures of all three angles?

31. The measures of two consecutive angles of a parallelogram are in a ratio of 4:5. Find the measure of the larger angle.

32. The perimeter of an isosceles triangle is 52 cm, and each leg is 10 cm less than the base. What are the lengths of the three sides of the triangle?

33. At 3 P.M., a 5½-ft-tall person casts a shadow of 8 ft. A nearby tree is 21 ft tall. To the nearest tenth of a foot, how long is the tree's shadow at 3 P.M.?

34. One of the legs of a right triangle is 17 in more than the other. Find the length of the larger leg, given that the hypotenuse is 25 in long.

35. A rigid cable is needed to support a 12-ft antenna on the roof of a building. The cable is to be attached to the top of the antenna and to a point 7 ft from the base of the antenna. To the nearest foot, how much cable must be available to accomplish this?

36. The parallel sides of an isosceles trapezoid are 16 and 30 cm. If the area of the trapezoid is 184 cm^2, how long, to the nearest tenth of a centimeter, are the nonparallel sides?

37. In the trapezoid described in the previous problem, what are the measures, to the nearest degree, of the four interior angles of the trapezoid?

38. A 15-ft ladder is leaning against a building at an angle of 39° with the ground. How high, to the nearest tenth of a foot, along the side of the building does the top of the ladder reach?

39. A floor design in the lobby of a building is in the shape of a rhombus, and its diagonals are laid out in red tile. There are as many floors in the building as there are whole degrees in the smaller angle of the rhombus. If the diagonals measure 6 and 8 m, how many floors are in the building?

40. From a point 25 ft from where a kite is directly over the ground, the angle of elevation of the kite is 32°. How long, to the nearest foot, is the string attached to the kite?

Problems Similar to Those in Chap. 5

41. Jennifer has been improving in math. Her last three test scores were in the ratio of 4:5:7. What was her lowest score if her test average is 72?

42. The weights of some of the members of the wrestling team are 172, 212, 185, 183, 201, and 179 lb. Find the median and mean weights of this group.

43. In an experiment, the times of many chemical reactions were recorded and are given in the table below. Find the mean, median, and mode of the reaction times.

Time, seconds	Number of reactions
23	31
25	46
28	52
31	61
35	10

44. Mike went to the video store to rent some newly released movies to watch over the weekend. Among the new releases there were four comedies, six action movies, and three dramas. If he wanted one of each, how many different possible groups of three movies could he rent?

45. Among the 12 students in the club, 3 have to be selected to be the officers. How many different groups of 3 students can be chosen to be officers?

46. Four cards are drawn from a standard deck of playing cards without replacing a card after it is drawn. If the first three cards are hearts, what is the probability that the next card drawn is not a heart?

47. "Rolling doubles" means that when two dice are tossed, the up sides are the same. In a certain game, you lose a turn whenever doubles are rolled. How many times should you expect to lose a turn if you toss the dice 78 times during the game?

48. On a circular spinner whose radius is 6 cm, the blue region occupies 9π cm^2, the green region occupies 12π cm^2, and the red and yellow regions equally occupy the rest of the area. What is the probability of the spinner stopping in the red region?

49. A group of seven boys and five girls went to the amusement park. Only four people were able to get on the last remaining car on the roller coaster. To the nearest hundredth, what is the probability that the four people consisted of two boys and two girls?

50. In a bag of cookies, there are four chocolate-chip cookies, eight sugar cookies, and six oatmeal cookies. Steve reaches into the bag without looking and takes out two cookies. What is the probability that they are both chocolate-chip cookies?

Answers to Chap. 6

1. $40.50.
2. 8 ft.
3. $1\frac{1}{3}$ mi.
4. Danny gained 3.4 lb.
5. 14 bowls of soup.
6. $4155.20.
7. Josh should wait because the discounted price on the helmet will be $42.50.
8. 64.

9. 21, 23, 25, and 27.
10. 8, 9, and 10.
11. $\frac{4}{7}$.
12. 47.
13. 2 years old.
14. 16 mph.
15. 15 seconds.
16. 35 lb.
17. 8 dimes.
18. 18 minutes.
19. $2\frac{1}{2}$ hours.
20. $3000 in the annuity and $7000 in the stocks.
21. 3850 students.
22. 60 and 84.
23. 12 and 3.
24. Gayle is 36 years old now.
25. 200 votes; Nicole received 1200 votes, Kristen received 1000 votes, and Lindsey received 400 votes.
26. $1\frac{1}{9}$ cups of flour.
27. Committee A grades 150 papers per hour; committee B grades 120 papers per hour. All the papers were graded in $2\frac{1}{3}$ hours.
28. 2.6°C.
29. 230 L.
30. 36°, 56°, and 88°.
31. 100°.
32. 14, 14, and 24 cm.
33. 30.5 ft.
34. 24 in.
35. 14 ft.
36. 10.6 cm.
37. 49°, 49°, 131°, 131°.
38. 9.4 ft.
39. 73 floors; the angle is approximately 73.7°.
40. 29 ft.
41. 54.
42. The median weight is 184 lb; the mean weight is 188.7 lb.
43. The mean is 27.8 seconds, the median is 28 seconds, and the mode is 31 seconds.
44. 72.
45. 220.

46. 39/49.

47. 13.

48. 5/24 or 0.208333 . . . .

49. 0.42

50. 12/306 or 2/51 or approximately 0.04.

A Brief Review of Solving Equations

The information below will refresh your memory about some of the most important facts about solving equations.

The two most common types of equations that you need to solve are *linear equations* and *quadratic equations*. For each there are several steps to follow that will ensure that you get the correct answer every time.

Linear Equations

An equation is *linear* if the highest power of the variable is 1 such as $4x - 5 = 11$. In fact, you won't even see the power since we rarely write exponents of 1. The goal is to undo all the complications in the equations and work your way to having the variable appear on only one side, with a numerical term on the other. The following example illustrates the necessary steps involved:

Example. Solve the equation $4(2x + 5) + 6x = 8x - 4 - 2x$ for the value of x.

Step 1 Use the distributive property to "remove" any parentheses in the equation.

$$8x + 20 + 6x = 8x - 4 - 2x$$
$$14x + 20 = 6x - 4$$

Step 2 Combine like terms on each side of the equation separately.

At this point the variable should appear at most once on each side, and a number should appear at most once on each side.

| **Step 3** | Remove the variable with the smaller coefficient by adding its signed opposite to both sides of the equation, and combine the variable terms. | $14x + (-6x) + 20 = 6x + (-6x) - 4$
 $8x + 20 = -4$ |

| **Step 4** | Remove the numerical term from the side of the equation with the variable by adding its signed opposite to both sides of the equation and combine the numerical terms. | $8x + 20 + (-20) = -4 + (-20)$
 $8x = -24$ |

| **Step 5** | Divide both sides of the equation by the coefficient of the variable and simplify. | $8x \div 8 = -24 \div 8$
 $x = -3$ |

Step 6 Check your answer in the original equation by using it in place of the variable.

$$4(2x + 5) + 6x = 8x - 4 - 2x$$
$$4(2 \times (-3) + 5) + 6 \times (-3) = 8 \times (-3) - 4 - 2 \times (-3)$$
$$4(-6 + 5) + -18 \qquad -24 - 4 + 6$$
$$4 \times (-1) + -18 \qquad -22$$
$$-4 + -18$$
$$-22 \qquad\qquad\qquad \checkmark$$

Example.   Solve the equation $4(y - 7) - 2(8 - y) = 6(y + 3) - 3y + 1$.

Step 1 Use the distributive property to "remove" any parentheses in the equation.

$$4y - 28 - 16 + 2y = 6y + 18 - 3y + 1$$

Step 2 Combine like terms on each side of the equation separately.

$$6y - 44 = 3y + 19$$

Step 3 Remove the variable with the smaller coefficient by adding its signed opposite to both sides of the equation, and combine the variable terms.

$$6y + (-3y) - 44 = 3y + (-3y) + 19$$
$$3y - 44 = 19$$

Step 4 Remove the numerical term from the side of the equation with the variable by adding its signed opposite to both sides of the equation, and combine the numerical terms.

$$3y - 44 + 44 = 19 + 44$$
$$3y = 63$$

Step 5 Divide both sides of the equation by the coefficient of the variable and simplify.

$$3y \div 3 = 63 \div 3$$
$$y = 21$$

Step 6 Check your answer in the original equation by using it in place of the variable.

(Try the check yourself. Each side has a value of 82 when $y = 21$.)

Key Fact: You can add, subtract, multiply, or divide only if you perform the same operation on *each side* of the equation.

Simultaneous Linear Equations with Two Variables

A pair of equations are said to be a *simultaneous* pair if you are told to look for the values of each variable that satisfy *both* of the equations. Only one pair of numbers will work. For example, the pair of numbers $x = 7$ and $y = 2$ is the only pair that will satisfy the equations $x + y = 9$ and $x - y = 5$ at the same time. We usually write the solution as an *ordered pair*, (x, y). In this case the solution is (7, 2). Simultaneous equations arise frequently in word problems that involve two unknowns. There are many methods for solving simultaneous pairs of equations. The two most common methods are by substitution and by combining.

Substitution is usually used when one equation lets you easily write one variable in terms of the other.

Example. Solve the following pair of simultaneous equations for x and y:

$$2x + y = 6 \qquad y - 3x = -14$$

Step 1 Solve one of the equations for one of the variables.

The second equation is easily solved for y by adding $3x$ to both sides, giving $y = 3x - 14$.

Step 2 Use the expression for the solved variable in its place in the other equation.

Substituting $3x - 14$ for y in the first equation gives $2x + 3x - 14 = 6$.

Step 3 Solve the "new" equation for its variable.

Combining and solving for x, we have
$$5x - 14 = 6$$
$$5x = 20$$
$$x = 4$$

Step 4 Use this variable to find the other variable from the equation you used in Step 1.

$$y - 3x = -14$$
$$y - 3(4) = -14$$
$$y - 12 = -14$$
$$y = -2$$

Step 5 Write your solution.	The solution is the pair $x = 4$ and $y = -2$ or, as an ordered pair (x, y), $(4, -2)$.
Step 6 Check that your solution satisfies both equations.	In the first equation $\quad 2(4) + (-2) = 6$ $\qquad\qquad\qquad\qquad 8 + (-2) \quad \checkmark$ $\qquad\qquad\qquad\qquad 6$ In the second equation $\quad -2 - 3(4) = -14$ $\qquad\qquad\qquad\qquad -2 - 12$ $\qquad\qquad\qquad\qquad -14 \qquad \checkmark$

Example. Solve the following pair of simultaneous equations for a and b:

$$3a - b = 20 \qquad 2a + b = 15$$

Step 1 Solve one of the equations for one of the variables.	The second equation is easily solved for b by adding $-2a$ to both sides, giving $b = -2a + 15$.
Step 2 Use the expression for the solved variable in its place in the other equation.	Substituting $-2a + 15$ for b in the first equation gives $3a - (-2a + 15) = 20$.
Step 3 Solve the "new" equation for its variable.	Combining and solving for x, we have $$5a - 15 = 20$$ $$5a = 35$$ $$a = 7$$
Step 4 Use this variable to find the other variable from the equation you used in step 1.	$$2a + b = 15$$ $$14 + b = 15$$ $$b = 1$$
Step 5 Write your solution.	The solution is the pair $a = 7$ and $b = 1$ or, as an ordered pair (a, b), $(7, 1)$.
Step 6 Check that your solution satisfies both equations.	In the first equation $\quad 3(7) - 1 = 20$ $\qquad\qquad\qquad\qquad 21 - 1$ $\qquad\qquad\qquad\qquad 20 \quad \checkmark$ In the second equation $\quad 2(7) + 1 = 15$ $\qquad\qquad\qquad\qquad 14 + 1$ $\qquad\qquad\qquad\qquad 15 \quad \checkmark$

Solving by *combining* is sometimes easier as it immediately eliminates one of the variables. Using the last example, we can add the given equations to eliminate b.

$$3a - b = 20$$
$$\underline{+2a + b = 15}$$
$$5a = 35$$

and we immediately see that $a = 7$. We can continue by using this value in one of the equations to find the value of the other variable as in step 4 above.

If the equations can't be added to eliminate one of the variables, we can "fix" the situation by doing a bit of multiplying first.

Example.   Solve the following pair of simultaneous equations for x and y:

$$3x + 2y = 36 \qquad 4x - 5y = 2$$

Step 1 Select one of the variables to eliminate and multiply each equation by the coefficient of this variable from the other equation.	If we choose to eliminate x, we will multiply the first equation by 4 and the second by 3, giving us the pair $12x + 8y = 144$ and $12x - 15y = 6$.
Step 2 Multiply both sides of one of these "new" equations by -1.	Multiplying both sides of the second equation by -1 gives us the pair $12x + 8y = 144$ and $-12x + 15y = -6$.
Step 3 Add the equations to eliminate the variable, and solve the resulting equation for the other variable.	Adding the two equations gives us $23y = 138$ and $y = 138 \div 23 = 6$.
Step 4 Use this value in either of the two original equations to find the value of the other.	Using $y = 6$ in the first equation, we have $3x + 12 = 36$, $3x = 24$, and $x = 8$.
Step 5 Write your solution.	The solution is the pair $x = 8$ and $y = 6$ or as an ordered pair (x, y), $(8, 6)$.
Step 6 Check that your solution satisfies both equations.	(Try the check yourself as practice.)

Quadratic Equations

In a *quadratic equation*, the highest power of the variable is 2. For example, the following are quadratic equations: $x^2 = 25$, $y^2 - 5y = 0$, and $n^2 - 6n + 8 = 0$.

Key Fact: Every quadratic equation can be solved for *two values* of the variable.

Example. $x^2 = 25$ has the obvious solutions $x = 5$ and $x = -5$.

Example. $y^2 - 5y = 0$ has the solutions $y = 0$ and $y = 5$.

159

Example. $n^2 - 6n + 8 = 0$ has the solutions $n = 2$ and $n = 4$. We sometimes find that these two solutions are equal.

Example. $x^2 - 8x + 16 = 0$ has the solutions $x = 4$ and $x = 4$.

The most common method of solving a quadratic equation is *by factoring* using the steps described below.

Example. Solve for all values of x that satisfy $x^2 = 25$.

Step 1 Rewrite the equation so that one side is 0.

This is accomplished by adding -25 to both sides: $x^2 + (-25) = 25 + (-25)$ or $x^2 - 25 = 0$.

Step 2 Factor the quadratic expression.

This particular quadratic is a difference of two perfect squares, which factors as $(x + 5)(x - 5) = 0$.

The next step uses the fact that if the product of two expressions is 0, then one of the factors must be 0. (How else could you multiply two numbers to get 0?)

Step 3 Set each factor equal to 0, and solve for the variable.

We have the two equations, $x + 5 = 0$ and $x - 5 = 0$, which have the solutions $x = -5$ and $x = +5$.

Step 4 Check each solution in the original equation.

Clearly, $x^2 = 25$ when x is $+5$ or -5. ✓

Example. Solve for all values of y that satisfy $y^2 - 5y = 0$.

Step 1 Rewrite the equation so that one side is 0.

This is already the case: $y^2 - 5y = 0$.

Step 2 Factor the quadratic expression.

Both terms of the particular quadratic have the common factor y, so it factors as $y(y - 5) = 0$.

Step 3 Set each factor equal to 0, and solve for the variable.

We have the two equations, $y = 0$ and $y - 5 = 0$, which have the solutions $y = 0$ and $y = +5$.

Step 4 Check each solution in the original equation.

Using $y = 0$, we have $\quad 0^2 - 0(5) = 0$
$$0 - 0$$
$$0 \quad ✓$$

Using $y = 5$, we have $\quad 5^2 - 5(5) = \quad 0$
$$25 - 25$$
$$0 \quad ✓$$

160

Example. Solve for all values of n that satisfy $n^2 + 8 = 6n$.

Step 1 Rewrite the equation so that one side is 0.   This is done by adding $-6n$ to both sides, giving us $n^2 - 6n + 8 = 0$.

Step 2 Factor the quadratic expression.   This particular quadratic factors as $(n - 2)(n - 4) = 0$.

Step 3 Set each factor equal to 0, and solve for the variable.   We have the two equations, $n - 2 = 0$ and $n - 4 = 0$, which have the solutions $n = +2$ and $n = +4$.

Step 4 Check each solution in the original equation.   Using $n = 2$, we have $2^2 + 8 = 6(2)$
                                        $4 + 8$ 12
                                           12 ✓
                  Using $n = 4$, we have $4^2 + 8 = 6(4)$
                                    $16 + 8$ 24
                                         24 ✓

Example. Solve for all values of x that satisfy $x^2 = 8x - 16$.

Step 1 Rewrite the equation so that one side is 0.   This is done by adding $-8x$ and $+16$ to both sides, giving us $x^2 - 8x + 16 = 0$.

Step 2 Factor the quadratic expression.   This particular quadratic factors as $(x - 4)(x - 4) = 0$.

Step 3 Set each factor equal to 0, and solve for the variable.   We have the two equations, $x - 4 = 0$ and $x - 4 = 0$, which have the equal solutions $x = 4$.

Step 4 Check each solution in the original equation.   Using $x = 4$, we have $4^2 = 8(4) - 16$
                                          16 $32 - 16$
                                            16 ✓

Key Fact: There are quadratic expressions that do not factor. To solve these, you need to apply the following special *quadratic formula*. The quadratic expression $ax^2 + bx + c = 0$ has the two solutions:

$$x = \frac{-b \pm \sqrt{b^2 - 4ac}}{2a}$$

(where a, b, and c represent the numbers that appear in the expression in those specific places; i.e., they are the *coefficients* of the terms).

This is beyond the scope of this book, but the following example will show how it can be used. The equation $x^2 - 9x - 22 = 0$ has the coefficients $a = 1, b = -9$, and $c = -22$. These values are substituted into the formula, and we have

$$x = \frac{-(-9) \pm \sqrt{(-9)^2 - 4(1)(-22)}}{2(1)}$$

$$= \frac{9 \pm \sqrt{81 + 88}}{2} = \frac{9 \pm \sqrt{169}}{2} = \frac{9 \pm 13}{2}$$

$$= \frac{9 + 13}{2} = \frac{22}{2} = 11 \qquad x = \frac{9 - 13}{2} = \frac{-4}{2} = -2$$

You should check these values to verify that they are solutions. Also, this quadratic, $x^2 - 9x - 22 = 0$, can be solved by factoring, and you should do so as practice.

Index

The index below includes the most relevant pages to the items listed. Within the statements of problems or their solutions, there may be ideas related to these items not included here.